P9-BAT-743

Through a reporter's eyes

The Life of
Stefan Banach

To My Parents

Roman Kałuża

Through a reporter's eyes

The Life of
Stefan Banach

Translated and edited by
Ann Kostant
Wojbor Woyczyński

Birkhäuser
Boston • Basel • Berlin

Author
Roman Kałuża
Filtrowa 77/38
02032 Warszawa
Poland

Editors and Translators
Ann Kostant
Birkhäuser
675 Massachusetts Avenue
Cambridge, MA 02139

Wojbor Woyczyński
Center for Stochastic and Chaotic Processes
in Science and Technology
Case Western Reserve University
Cleveland, Ohio 44106

Library of Congress Cataloging-in-Publication Data
Kałuża, Roman
   [Stefan Banach. English]
   Through a reporter's eyes : the life of Stefan Banach / Roman
Kałuża ; translated and edited [by] Ann Kostant, Wojbor Woyczyński.
     p.   cm.
   Includes bibliographical references and index.
   ISBN 0-8176-3772-9 (h : acid-free). -- ISBN 3-7643-3772-9 (h :
acid-free)
   1. Banach, Stefan, 1892-1945. 2. Mathematicians--Poland-
-Biography. I. Kostant, Ann, 1937- . II. Woyczyński, W. A.
(Wojbor Andrzej), 1943- . III. Title.
QA29.B33K3513 1995   95-25811
510'.92--dc20   CIP
[B]

Printed on acid-free paper
© 1996 Birkhäuser Boston

Copyright is not claimed for works of U.S. Government employees.
All rights reserved. No part of this publication may be reproduced, stored in a retrieval system, or transmitted, in any form or by any means, electronic, mechanical, photocopying, recording, or otherwise, without prior permission of the copyright owner.
Permission to photocopy for internal or personal use of specific clients is granted by Birkhäuser Boston for libraries and other users registered with the Copyright Clearance Center (CCC), provided that the base fee of $6.00 per copy, plus $0.20 per page is paid directly to CCC, 222 Rosewood Drive, Danvers, MA 01923, U.S.A. Special requests should be addressed directly to Birkhäuser Boston, 675 Massachusetts Avenue, Cambridge, MA 02139, U.S.A.

ISBN 0-8176-3772-9
ISBN 3-7643-3772-9

Typeset by Martin Stock, Cambridge, MA 02139
Printed and bound by Maple-Vail, York, PA
Printed in the United States of America
9 8 7 6 5 4 3 2 1

Contents

Foreword to the Polish Edition

SO FAR, SURPRISINGLY LITTLE has been written about the life of
Stefan Banach— one of the founders of functional analysis and one of
the most original mathematicians of the 20th century. A small brochure
of less than a score of pages written by Hugo Steinhaus and descriptions
of small episodes dispersed in reminiscences of various mathematicians
(Kazimierz Kuratowski, Stanisław Ulam and others) are all that has
existed until now.

This scarcity is a fact worth contemplating. After all, Stefan Banach
is one of the mathematical world's most recognizable names. In spite
of this, an average Polish intellectual knows few hard facts about his
life. Andrzej Alexiewicz (deceased in July 1995), a Professor at Poznań
University who knew Banach well, claimed that many foreign mathematicians are not even aware that Banach was Polish. One edition of the
Encyclopedia Britannica has him listed as a Russian.

Several years ago, the German publisher Teubner Verlag asked Inter-
press Publishers to prepare materials for a first biography about Stefan
Banach. Finding an author in the mathematical community turned out to
be impossible. After much hesitation, I undertook to prepare the monograph. The process did not go very smoothly, and twice before the book
was finished I nearly gave up on this enormous responsibility and very
time-consuming work.

Some explanations are in order:

I am not a mathematician. At the beginning of the writing process I
tried to secure the collaboration of a scholar working in that discipline.
Nineteen mathematicians refused my offer and two others warned me
against undertaking this task. Thus I was forced to limit myself to the

concept of a mathematician's biography from a reporter's perspective which only to a very modest degree analyzes his scientific work.

The present text contains many pieces of information that have never been published. This comment applies in particular to Chapters I–IV and VII. Chapter VI, *Banach Privately and in Daily Life*, is a synthesis of dozens of oral interviews (which I conducted with the help of Maciej Maniowski) with his students, friends, and collaborators, and with Stefan Banach's half-sister Antonina Waksmundzka (née Greczek). Sometimes the revelations of people who knew Banach were mutually contradictory. Usually in such situations it is the author's responsibility to evaluate the validity of different claims, but I thought it appropriate not to dot my i's. In any case, we often had too little information about Banach to pass judgment.

The fact that the book happens to contain little information of strictly mathematical character makes it more accessible to a broadly educated audience; my goal was to write a popular scientific biography. However, in Chapter VIII, I permitted myself to include excerpts from writings of mathematicians from different countries (Poland, Hungary, the former Soviet Union and the US) containing opinions about Banach's work from a mathematical perspective. With the help of Stanisław Kwapień of Warsaw University, I included short commentaries on the contents of a number of Stefan Banach's papers, describing their significance and role in the development of mathematics.

The reader may wonder why the density of hard facts decreases in the text following the first two chapters. The reason is that the information about certain periods of Banach's life (for example the years 1911–1913 and 1939–1945) is very scarce. There are no documents (and I went through a score of Polish archives), most of the people who were eye witnesses have died, and the passage of time has dimmed the memories of others.

It is my hope that after this book appears, those whom I was unable to reach, either by repeated appeals in the press or via private contacts, will surface. I also count on critical voices and commentaries which, sometime in the future, would help me to prepare a more complete biography of one of the greatest scientists of the 20th century.

Finally, I would like to thank the people without whom this work could not appear: the referees, the late Professor Andrzej Alexiewicz and editor Maciej Iłowiecki for comments which led to many improvements; Professor Stanisław Kwapień for mathematical consultations, for brief descriptions of specific Banach publications, and the translation of their foreign language titles; Maciej Maniowski for help in collecting the interviews; and above all the untiring editor of this book, Ms. Ewa Trzeciak.

Warsaw, June 1980

Translator and Editor's Preface

THE YEAR 1992 MARKED the 100th anniversary of Stefan Banach's birth. In Europe, and in particular in his native Poland, the event was commemorated by several conferences and symposia organized by the mathematics community and by other official pageantry. And yet, although half a century has passed since his untimely death, the general English-speaking public has had until now no access to a more in-depth life story of a mathematician whose name is one of those most often encountered in modern mathematical research writings. This small volume is an effort to fill that gap in the biographical literature.

Its author, Roman Kałuża, is a reporter and journalist well known in Poland, especially for his work as longtime scientific editor of the once influential weekly *Kultura*. He is not a professional mathematician, but his interest in this icon of Polish mathematics was piqued by the fact that there was so little written about Banach for the general Polish public. Banach was put on a pedestal and left there, with all kinds of legends swirling around his, by now, half-mythical figure.

From the inception of the original idea, it took some ten years until the book was finally published in Poland in 1992, although the manuscript was ready several years earlier. The 1980s were not exactly quiet times in that part of Europe, and the turbulence in both local and global politics certainly did not help the publication process. Also, not everybody was keen on the idea of taking Banach off his pedestal. However, the Banach that emerges from the book's pages is a sympathetic, warmly presented, thoroughly human, and obviously fascinating genius, anything but a stilted, traditional, European-style university professor and ivory-tower academician.

The book is not a scholarly treatise – the type that would be written by an historian of science. Perhaps such a text still waits to be written, but Kałuża did some solid archival work and the result is a unique biography. The journalistic provenance, with its emphasis on interviews and verbatim quotation of sources, is obvious, so we decided to mention it explicitly in the subtitle (*Through a Reporter's Eyes*).

The idea of an English translation came to one of us (W.A.W.) from Stanisław Kwapień, Professor of Mathematics at Warsaw University, who originally helped Kałuża by interpreting some of Banach's abstract mathematics. After some consideration we thought that it would be a good idea, and undertook extensive reorganizing, editing, and rewriting (with the author's benevolent permission) as well as the substantial addition of new relevant material to address a more universal audience. Photographs and pictorial material were added, mostly from the private collection of W.A.W., as the original edition had none. An index of names was added, and one of us (W.A.W.) wrote a new and extended appendix on mathematics in Banach's times.

Finally we would also like to thank Stan Kwapień, Johanna Holbrook, Irene Dennis, and Asha Weinstein for reading and commenting on portions of this book, Martin Stock for his book design and typography, Arnold Wedel for his helpful contributions, and Alek Weron for providing the text of Kazimierz Szałajko's inaugural address.

A.K. and W.A.W.
Boston and Cleveland,
on the 50th anniversary of Banach's death,
August 31, 1995

CHAPTER I

Beginnings and Premonitions

ONE OF THE GREAT MATHEMATICIANS of the twentieth century, Stefan Banach is associated with a wide range of fundamental and original mathematics, particularly in the area of functional analysis. In Poland, he is considered a national hero.

Banach was born on Wednesday, March 30, 1892 in Cracow at St. Lazarus General Hospital.[1] The names of the primary and secondary doctors on duty in the ward, as well as the midwife who probably delivered the child are recorded in hospital documents.[2] The birth certificate states that Banach's mother was Katarzyna Banach and that his father was Stefan Greczek, a [low-level] civil servant. They were not married.

Banach never knew his mother,[3] who gave him up after his baptism into the Roman Catholic Parish of St. Nicolaus on April 3, 1892. On many occasions Banach tried to learn something about her from his father, but the mystery was never solved since his father refused to reveal her identity or divulge any information whatsoever on the subject. Greczek explained that he was under oath to keep the secret. He later married twice, and had one son from the first marriage and four children by his second wife.

According to Hugo Steinhaus, Banach's mentor and later close friend and mathematical colleague,[4] Stefan Greczek was born into a family of mountaineers in the highlands area of the Jordanów district, and worked as a civil servant in the headquarters of the Cracow Railroads. But this is not true. From documents preserved by the Austrian Imperial-Royal Main Tax Office in Cracow,[5] we know that Greczek was employed as a tax adjunct [perhaps some sort of aide] beginning in 1903 (earlier dates are not documented) in the Tax Office at 14 Graniczna Street. Two years later, in 1905, the Official Lvov Gazette[6] notes that he received for his services a Silver Cross of Merit and was promoted to the post of tax official.

Additional knowledge about Greczek comes from his daughter from his second marriage, Antonina Waksmundzka. She said that her father was born in 1868 in the village of Ostrowsko, which is outside the provincial town of Nowy Targ,[7] and that he died in 1968 at the age of 100, having outlived his son Stefan by almost a quarter of a century. Stefan Greczek's father Józef had been a farmer and village mayor. Józef's wife, Antonina (née Borkowska), was reported to have been a countess, or at least something of an aristocrat, since her family bore the coat-of-arms of the Pomian clan. Whatever her lineage, the Greczeks of Ostrowsko were quite poor.

As a teenager, Stefan Greczek was forced to emigrate to Hungary to find a job – which he did – in a local brickyard in Pest. He was bright and alert. From letters he wrote to his family we know that, by perfecting his German, he was assigned within a short time to duties in the company's accounting office, where he acquired some practical skills. We also know that Greczek at some point served as a noncommissioned officer in the Austrian Army, and that he was given a post in the office of the 20th Imperial-Royal Infantry Regiment stationed in Cracow.

Family legend has it that Banach's early childhood was spent with his grandmother in Ostrowsko. Whether or not this is true, young Stefan is known to have been very attached to her, feeling more affection for her than for any of his other relatives. When she became ill, his father sent him to Cracow and entrusted his future upbringing to two women,

Franciszka Płowa and her daughter Maria. Stefan continued to visit his grandmother often until she died; he was present at her funeral.

By the standards of those days, Banach lived in relative comfort with the Płowas, who were apparently well off. Franciszka's husband was the manager of the Krakowski Hotel, located in the lovely and fashionable Planty[8] district, while Franciszka herself worked in one of the five branches of the "Tęcza" [Rainbow] laundries. But despite the warmth he felt for Franciszka Płowa, regarding her as his foster-mother and Maria as his big sister, Banach's memories of childhood were not all that happy.

The first person to recognize Banach's talents was Juliusz Mien, an intellectual of French origin who arrived in Cracow in 1870. Mien earned his living as a photographer and translator of Polish literature, but he was also the guardian of Maria through whom he came to know Banach. Mien took the young boy under his wing, taught him French, and in general supervised his education – all without remuneration. Largely due to him, Banach achieved a fluency in French which greatly impressed his foreign colleagues at future international congresses. Mien probably fostered Stefan's early mathematical inclinations as well.

Several sources confirm that Stefan Greczek never forgot his son, even when he had established his own legitimate family. Not only did he often provide some financial help, but he maintained close contact with the Płowas who supervised and guided Banach's upbringing through his college years. That Greczek was married and financially stable was a matter of some importance in bourgeois Cracow. Helena, his first wife, bore him a son Wilhelm. Years later Wilhelm became an officer in the 21st Artillery Regiment, then a high ranking official, and finally president of the State Railroads in Cracow and Przemyśl. Perhaps Wilhelm is the reason for Steinhaus and others mistakenly thinking that Banach's father was a railroad official.

Contacts between father and son were apparently polite and cordial. In a postcard sent by Banach to his father, the gymnasium[9] student thanks Greczek for giving him money for an excursion to Kielce and the mountains north of Cracow. Banach's half-sister Antonina said that although Banach loved his father, he did not show him much warmth or filial affection. Remembering her father and half-brother, Antonina said:

> Physically they were very much alike, both broad-shouldered, tall and handsome...good-looking male specimens. My father, like Banach, was self-taught. Father's German was perfect; he spoke and read fluently in that language. In our home library – we lived then at Bonerowska Street in Cracow – there were works by Goethe, Schiller, and Heine. Father also knew Latin quite well and was very interested in politics. Father was a straightforward person, yet somewhat naïve. Although he was a bookkeeper, he did not approach life in a calculated fashion, but rather was given to grand gestures. For example, after saving a considerable number of gold ducats, which were later withdrawn from circulation by the Austrian authorities, he decided at some point to donate all of the coins to help restore the Dominican Church in Cracow. Deeply religious, it must have been something of a tragedy for him to divorce his first wife Helena and marry my mother, Albina Adamska. From that second marriage were born my brothers: Kazimierz, a lawyer who was killed in 1939, Tadeusz, a physician living in the United States, Bolesław, who died when he was a second-year law student, and myself, a pharmacologist by education. But that of course was much later.

Few documents have been preserved from the first stages of Banach's education. However, numerous materials from his gymnasium years attest to the fact that this period was instrumental in shaping both his character and mathematical proclivities.

In 1902, after finishing elementary school, Banach, then 10 years of age, entered the first grade at Cracow's Henryk Sienkiewicz Gymnasium Number 4. The gymnasium, which specialized in the humanities, was located at the Podwale [literally under the walls of the city] and was usually referred to as "at Goetz" because the building itself was leased from Cracow brewery owner Goetz-Okocimski.[10]

Banach's classmates and best friends were Witold Wilkosz, a future mathematician, and Marian Albiński. Albiński's memoirs provide us with the only relatively trustworthy characterization of Banach during this period. He and Banach attended the same gymnasium from 1902 to 1906, when Marian transferred to the [King Jan] Sobieski Gymnasium. The reasons for this change shed light on one aspect of the school system in the Austrian part of partitioned Poland. Albiński wrote that the cause of his transfer was a conflict with his Greek teacher. When at the end of the first semester he received an "F" in Greek, the failing grade invoked a stiff penalty in the form of a stamp-fee amounting to 20 crowns. Here is what Albiński in his memoirs wrote about his best friends:

> Wilkosz also moved on with me to the Sobieski Gymnasium; I do not know what his reasons were. Banach remained in Gymnasium Number 4, until passing the *matura*[11] examination in 1910.
>
> After I left the gymnasium "at Goetz," my contact with Banach diminished, but Wilkosz managed to maintain close contact with him. Since I was friendly with Wilkosz, I saw both of them together on numerous occasions.
>
> Stefan Banach, as I remember him, was a good friend. Quiet, but not without a gentle sense of humor, he had a rather secretive nature. He always wore a clean and neat uniform, as we all did. His financial situation compelled him to tutor younger students as well as students "downtown," but he never seemed needy. I should add that he tutored his own classmates without pay.
>
> From the early grades on Banach and Wilkosz were attached to each other through their love for mathematics. During school breaks I would often see them working on solutions to mathematical problems which, for me, a humanist, might as well have been Chinese.
>
> Banach's friendship with Wilkosz was not restricted to the school grounds; they met after school in Wilkosz' home on Zwierzyniecka Street, in the school building, or at the Planty Krakowskie. In later years, if a mathematical question was bothering them, the two friends engrossed in thought would keep walking each other home through the streets of Cracow for half the night.
>
> I never took part in those mathematical discussions, but I had long conversations with Wilkosz, with whom I remained close. What connected us in the gymnasium and after the matura examination were our literary interests and a weakness for the same female students.

After the matura, Wilkosz graduated in mathematics from the Jagiellonian University and went on to become a professor of mathematics there.

There are documents to the effect that Banach was a very diligent student. It should be noted that the educational program of the gymnasium at that time emphasized classical subjects – Latin, Greek, modern

5

languages, history, etc. – over mathematics and the natural sciences. Banach, under this typical program, did not follow a course that was entirely suitable to his predisposition and interests. All too often he was taught mathematics by rather incompetent people. In reminiscing about that period, Banach did not speak very kindly about the quality and methods of instruction in his favorite subject.

Numerous documents exist concerning Banach's attendance in the second grade of the Imperial-Royal Gymnasium Number 4. For today's many school reformists, it may be interesting as well as instructive to examine the gymnasium curriculum of that period: Religion (two hours per week); Latin (eight hours per week); Polish (three hours per week); German (five hours per week); History and Geography (four hours per week); Mathematics (three hours per week); Natural history (two hours per week).[12] There were also extracurricular subjects: history of the native country, French (according to records, nobody in the second grade attended that class), singing, drawing, calligraphy, gymnastics, and stenography.

Mathematics in the second grade was taught from the text *Beginnings of Arithmetic and Algebra* by Brzostowin, and from the book *Geometry and the Imagination* by Mocnik-Maryniak.

The teaching faculty was small. Thanks to the efforts of the gymnasium principal, Antoni Pozdrowski, a branch of the Nowodworskie Gymnasium had been transformed into the independent Gymnasium Number 4 just before Banach began his studies there. Pozdrowski was on the board of the Sixth Imperial-Royal School Councilor's office and taught mathematics. Known for his strict discipline, he must have exerted a strong influence on Banach's mathematics at that time.

Other teachers were:

– Ludwik Boratyński, Ph.D. in Philosophy and Curator of teaching materials for history and geography, who taught both of these subjects for all grades;

– Roman Gutwiński, professor of the VIIIth rank,[13] member of the Physiographical Commission of the Cracow Academy of Knowledge and Curator of the natural history room, who taught natural history and mathematics;

– Aleksander Stiasny, homeroom teacher, who instructed in Latin, history, and geography;

– Father Paweł Pyłko, Ph.D. in Theology, who taught religion;

– Antoni Lekszycki, who taught Polish.

Later arrivals to the faculty included Kamil Kraft, Ph.D. in All-Medical Sciences, who taught mathematics and physics, and there were others as well.[14] Many years later, upon receiving the Scientific Prize of the City of Lvov, Banach acknowledged that his mathematical interests had been fully awakened and formed by Kamil Kraft.

The gymnasium attended by Banach was not particularly exclusive, but like others in Cracow it maintained very strong ties with higher level scientific institutions, such as the university and Polish Academy of Knowledge. That academicians quite often worked as gymnasium teachers guaranteed a high level of secondary education, as can be seen from the above curriculum and faculty list at Banach's school. The best known and most exclusive gymnasiums of Cracow were St. Ann's Gymnasium (Nowodworskie), directed by Leon Kulczyński, Docent of Pedagogy at the Jagiellonian University, and Sobieski Gymnasium, run by Tomasz Sołtysik. The academic credentials of the teachers here were impeccable; usually they were alumni of the Jagiellonian University or of the Jan Kazimierz University in Lvov.

The Cracow historian and writer Karol Estreicher noted that the patriotism and attachment of the teachers to Polish culture were beyond question in both gymnasiums, but that the Sobieski suffered from shortcomings common to the system under the Habsburgs. Students there were not adequately prepared for study at the university in the areas of natural sciences and engineering since too much weight had been attached to mastering classical subjects.

Compared to the Sobieski or to St. Ann's, Banach's gymnasium was not very fashionable, but in retrospect this may have been for the best, since in this decisively formative period Banach avoided the negative experiences of his contemporaries in the two other gymnasiums. A description of the prevailing atmosphere in these institutions is provided by Estreicher.[15] The Sobieski Gymnasium was attended not only by sons of the Cracow intelligentsia but by the youngest offspring of Polish aristocrats and larger landowners: the Tarnowskis, Potockis, Sapiehas, Lubomirskis, Woronieckis, and the Zaleskis. The presence of these upper-class children sometimes had a paralyzing effect on pupils of more modest background, bringing out emotions ranging from inferiority to vanity and a desire to shine at all costs. Leon Chwistek, a well-known logician-mathematician who was also a philosopher, painter, and writer, later blamed the psychological immaturity that sometimes handicapped him on his gymnasium experience:

> In Poland, membership in the gentry was absolutely obligatory if you wanted to be somebody. People with names like Korzeń (Root), Wrona (Blackbird) or Kurczątko (Chicken), even when such names sounded French, had absolutely no chance of qualifying for "genius" status. When it happened that a merit certificate had to be awarded to a certain Trzęsigruszka (literally Shakes-*pear*), all of a sudden it had to be demonstrated that his name was not Trzęsigruszka after all but Trzęsiwłócznia (Shake-*spear*) – [a pun not too awful in translation], which was more in keeping with "upper-class" lineage.

Chwistek's comments were published in the *Pion* literary magazine (1935) and quoted in a biography of him by Estreicher, who claimed that these assertions were exaggerated and further evidenced Chwistek's acute inferiority complex. Nevertheless it was true that students of landowning and aristocratic families had a tendency to flaunt their superiority, their connections, and the brilliant careers awaiting them. Estreicher also says that the Sobieski Gymnasium's principal and the teachers reporting to him favored those pupils of the "upper classes" and listened attentively to their well-connected parents. What is curious about the sense of inferiority expressed by Chwistek and some of his fellow alumni[16] is that all of them were born into old, well known, and respected families.

In contrast with the Sobieski Gymnasium, the environment at Gymnasium Number 4 was less intimidating. Part of the student community was Jewish and part was from the same social circle as Banach. None of his classmates came from conservative Cracow's upper crust. In those days this factor could not be dismissed and certainly had an influence on Banach's later political leanings and rapport with others. Listed in his class are such unaristocratic names as Piotr Owca (sheep), Stanisław Klocek (block), Albin Kawaler (bachelor), Stanisław Nosek (little nose), and Sikora (titmouse).

During his first years at the gymnasium Banach was classified as either of the first class (degree) with distinction, or of the first class (roughly meaning with highest honors); such honors were awarded to only two or three students a year, and many had to retake exams after the summer break.

Banach's grades were usually "excellent" in mathematics and natural sciences, while in other subjects he received "very good" and "good" grades.[17] Albiński remarks:

> I do not remember him ever receiving a "sufficient" grade. His modesty, or perhaps rather shyness, resulted in his not being very visible in the class picture from grade II which I preserved. In the photograph he is sitting at the third row desk and is partially obscured. At that time Banach's regular seat was actually in the first row, but when the photographer showed up to take class pictures, Banach quickly took advantage of the commotion and moved to the third row desk, occupying a seat next to mine.

In his memoirs Albiński also quotes a description of Banach by Adolf Rożek, a classmate who became a historian in Konstanz on Bodensee:

> Banach was slim and pale, with blue eyes. He was pleasant in dealings with his colleagues, but outside of mathematics he was not interested in anything. If he spoke at all, he would speak very rapidly, as rapidly as he thought mathematically. He had such an incredible gift for fast thinking and computing that his intelocutors had the impression that he was clairvoyant.
>
> Wilkosz was a similar phenomenon. Between the two of them, there was no mathematical problem that they could not speedily tackle. Also, while Banach was faster in mathematical problems, Wilkosz was phenomenally fast in solving problems in physics, which were of no interest to Banach.

There are interesting recollections concerning contacts between Banach and the school priest, Father Pyłko. It seems that Banach would often ask the priest such questions as "If the Lord is omnipotent, can he create a rock that he cannot lift himself?" Clearly Banach was already a skeptic. In a group photograph from those days, one can see a calm, slightly sarcastic, yet somewhat timid expression on his face. It was an expression that lasted until his final year in the gymnasium, when his enthusiasm for studying may have been dampened by an overload of work. From the age of 15 he tutored in mathematics, thus helping to offset the significant cost of his gymnasium education.

In addition to mathematics, Banach was also fond of Latin. Once he told Professor Andrzej Turowicz, a Dominican friar and later professor of mathematics at the Jagiellonian University, that "mathematics is too sharp a tool to put into the hands of children; for training in logical thinking, there is nothing better than *accusativus cum infinitivo* and *ablativus absolutus.*" As for his own gymnasium mathematics program, Banach commented, "It reminded me of using a bayonet for a toothpick." To be sure, Banach himself had mastered it inside and out, as well as a pretty large chunk of higher mathematics.

In 1910 Banach faced the matura along with other serious hurdles: the good student of earlier gymnasium years now had eight failing grades. Even the mathematics teacher despaired of getting him past the examining committee, despite his efforts to convince them that they were dealing with a real mathematical genius. However, the decisive vote was cast in Banach's favor by the kind and tolerant religion teacher, Father Pyłko, despite the skeptical student's frequently embarassing questions. The influence of the Catholic clergy was extremely powerful in elementary and secondary education until World War II.

Among the 27 pupils of the graduating gymnasium class, six were honored as "gifted with distinction." Banach was not among them, and had to be satisfied with being only one of the "gifted" students. Two students had to take makeup exams.

It is stated in the school chronicle that:

1. The 1909/10 school year commenced on September 3 with a solemn mass in St. Ann's Church, attended by the student body and the teaching faculty....

6. On January 30, 1910, students were issued their first semester grades....

9. The written matura examination was held on May 17, 18 and 19, and the oral examination on June 9–14 and was chaired by His Highness Sir Emanuel Dworski, Imperial-Royal Court Councilor and the Inspector of National Gymnasia. . . .

10. On June 28 a funeral service [*requiescat in pace*] in memory of the late Emperor Ferdinand was held and was attended by the teaching faculty and the whole student body.

11. On June 30, after solemn singing of the *Te Deum* and of the popular national anthem, the end of the year grade certificates were distributed and the school year was closed.

From the records of the gymnasium management, among the graduates deemed "mature" [by passing the matura examination], the following enrolled in colleges: two majored in theology, eight in law, three in medicine, six in philosophy, one in agriculture, three in engineering, two in trade, and one went to a military academy. Among the alumni headed for engineering studies was Stefan Banach.

CHAPTER II

The "Greatest Discovery" of Hugo Steinhaus

LITTLE IS KNOWN ABOUT the period of Banach's life immediately following his matura; people who knew him at that time are no longer alive, and many documents from those years were not preserved – or the author was unable to discover them. However, subsequent events prove that those years were important ones.

After the matura in 1910, Stefan Banach and his friend Witold Wilkosz discussed plans for their future. They concluded that since mathematics was so highly developed, it would be impossible to do anything new in that discipline, so majoring in mathematics was not advisable. Banach opted for an engineering program, while Wilkosz chose Oriental languages. The two parted. Later on Banach playfully admitted that he and Wilkosz, in their youthful naïveté, had been mistaken about the opportunities for doing original work in mathematics.

Banach's graduation and successful passing of the matura are said to have elicited from his father the comment: "I promised your mother that I would help you to get the matura certificate. From now on you are on your own."[1] He added some further advice, according to Banach's half-sister Antonina, when Stefan was preparing to leave for the Lvov Polytechnic: "Take good care of yourself and avoid diseases." We know that Banach, neither depressed nor offended by these comments, left Cracow without a penny to his name, and without the expectation of further financial help from his father.

For Wilkosz, who came from a respectable Cracow family, the situation was different. His father was a gymnasium teacher. Young Witold's talents were recognized at an early age. He was both linguistically inclined and mathematically gifted, as well as very industrious and entrepreneurial. In the fourth grade of the gymnasium, he had begun to study higher mathematics, and in the sixth grade commenced studying both Western and Eastern languages. As a gymnasium senior, he had written a scholarly paper on the basis of which he was admitted as a member of the Morgenländische Gesellschaft (Oriental Society), from which he obtained a stipend permitting him to spend a few months in Beirut. At the Jagiellonian University he initially enrolled in the classical philology program, but switched after two years to mathematics. Then, one year later, he left for Turin to further his studies in mathematics.

When World War I broke out Wilkosz accepted a position as a gymnasium teacher in the Silesian city of Zawiercie, and at the same time continued his scholarly pursuits. He received his Doctor of Philosophy degree from the Jagiellonian University, and in 1919, began lecturing there on an *ad hoc* basis; in 1920 he obtained his habilitation.[2] Within a few years he became a Contract Professor and then Professor Extraordinarius.[3] Highly respected as a teacher, he also popularized science and authored numerous technical papers and academic textbooks.

By contrast, Banach had to earn money to support his studies. Very likely he tutored. Therefore it is not surprising that it took him until 1914 to pass his half-diploma examinations (which means he completed his freshman and sophomore years) at the Lvov Polytechnic.

* * *

Lvov in those days was an important center of Polish culture and science. Those who lived there agree that it was a beautiful and unusual city. Of its 200,000 inhabitants, half the population was Roman Catholic, almost a third Jewish, and a significant number were Greek Catholics. In fact, there were three archbishops in Lvov, representing Roman, Greek, and Armenian branches of Catholicism. In this mixture of denominations, nationalities, and cultures, trade and industry developed rapidly, and art and science began to blossom in an unprecedented manner, especially in the natural sciences.

In the memoirs of a professor of that period, there is a description of the wondrous city of Lvov.

> Lvov was especially pretty. Its picturesque vistas reminded me of Prague, with its twisting and undulating streets, and its crowning view of the 13th century High Castle on the hills; farther away on another hill, Łyczaków, a dramatic, internationally unique cemetery; and higher up still, Stryjski Park, known for its Fall Fair and Exhibition. The whole city was dotted with old, architecturally pleasing buildings and churches... and churches... and more churches. And there, among all these breathtaking surroundings, men and women of Lvov, somehow very different from citizens of other cities – more benevolent, with sunny dispositions, often singing, as if everybody carried in his briefcase a supply of joy and smiles.
>
> I board a streetcar and ask for the High Castle. The conductor replies, grinning from ear-to ear:
>
> "You must be from Warsaw, aren't you? Then you must climb to the very top of the hill. You will see how wonderful our city is. Have you seen the cathedral? And the Boims Chapel? How about the Łyczakowski Cemetery? You ought to see them all!" At the right tram stop the conductor gets out of the car, points out the Castle to me, offers some friendly advice, and bids me farewell with the admonition: "You must fall in love with Lvov!"
>
> On another occasion, hurrying down the street I bump into a passer-by. I slow down and apologize in embarrassment. "Never mind Madam! I didn't even notice," says the stranger, and in a friendly gesture gives me a flower from his bouquet. Why? For no particular reason. He simply goes through life with a joyful smile on his lips.[4]

Social life blossomed in numerous cafés and wonderful Viennese-style pastry shops, among which the following were recommended by a guide book: Zalewski's at 22 Akademicka Street, Welz's at 5 Akademicka Street, Dudek's on Mariacki Square, Europejska at 5 Hetmańska

Street, Urbanik's at 35 Sykstuska Street and Lewartowski's at 6 Bielaw-skiego Street. They were all mainly located in the neighborhood of the University and the Polytechnic and were patronized by large numbers of students and faculty.

The main building of the Lvov Polytechnic was built between 1873 and 1887 and was located at 12 Leon Sapieha Street. The professor describes the campus.

> Surrounded by a wide expanse of lawn, the building, with its beauti-ful dark columns, glistened in the distance as if bathed in blood at sunset. On the façade there was the honorific Cross of the *Virtuti Militari* Order.
>
> Behind it, on Karpiński Street, there was an ugly chemistry build-ing, and to the right, on Nikorowicz Street, a modern library with a moving inscription by Kazimierz Bartel, a professor of descriptive ge-ometry and later prime minister of Poland: *Hic moritui et muti loquuntur* (here the dead and the mute speak). Further on, behind the same fence, stretched monastery buildings and the Mary Magdalene Church. At some point the monastery was converted into a prison for women, but later the Polytechnic converted the prison (which by that time had ac-quired the nickname "Little Magdalenes") into scientific laboratories. There, on the first floor, was the Department of Physical Chemistry. Thus I landed on the territory of a prison for incorrigible women, which my colleagues jokingly said was an appropriate place for an incorrigible female physico-chemical sinner.
>
> Compared to the Warsaw Department, this one was small, but the main lab could accommodate without any inconvenience 80 undergrad-uate students and a few graduate students working on their theses. The five-foot thick monastery outer walls created a serious and austere am-biance. My office was separated from the church nave by only a thin partition. In May, when the buzz of student activity abated, sitting at my desk in the evening, I virtually became a participant in the daily vespers with its magic of enchanting hymns.[5]

Several distinguished scholars taught at the Faculty of Engineering as well as two well-known politicians – the above-mentioned Bartel and Ignacy Mościcki, a chemical engineer and later President of Poland from 1926 to 1939. From the Polytechnic's archives we learn that the fresh-man curriculum at the Faculty of Engineering where Banach enrolled consisted of mathematics (taught by Placyd Dziwiński), descriptive ge-ometry and technical drafting, general mechanics, geology, chemical engineering, social economics, commerce, and financial, state, and administrative law.

The second-year curriculum consisted of mathematics (taught by Zdzisław Krygowski), general theoretical physics, engineering mechanics, metallurgy and wood technology, civil engineering, geodesic surveying, and lab courses in industrial and electrical engineering.

All we know of Banach at this time are the details of his academic record and that he was probably tutoring to earn a living. Little is known about his interests, friends, or how he spent time during his student days.

In the late spring of 1914 Banach passed the half-diploma examination before the Imperial-Royal Commission for State Examinations. In the Engineering Faculty, the Commission was chaired by Jan Bogucki, and included among others the mathematician Placyd Dziwiński, and physicist Kazimierz Olearski.

* * *

In July of 1914 World War I broke out. Russian troops began their offensive, and Banach left Lvov. Banach was excused from military service on two accounts: he was left-handed and had poor vision in his left eye. Turowicz says that Banach found a job as a foreman building roads somewhere in Galicia. Also, for some time during the war he lived in Cracow (he also periodically returned there between 1910 and 1914), but the exact dates are unknown.

We do know that in Cracow Banach initially supported himself by tutoring in all subjects at the gymnasium. The head mistress of a boarding school for young ladies offered him a regular position on her staff, but Banach turned it down. He had no interest in teaching high school. Perhaps by then he was thinking seriously about a research career in mathematics.[6]

As stated above, very little information about Banach exists for this period during the war.[7] No documents were found, even in the records of the Jagiellonian University, although it is clear that he attended some lectures there. The Jagiellonian University in Cracow, established in 1363 by King Casimir the Great, has been for centuries one of the great centers of learning in Central Europe. Since 1900, the dominant figure in mathematical circles there was Stanisław Zaremba (1863–1942), who embodied some of the best European scientific traditions. Born in Romanówka in the Ukraine, he was a graduate of the St. Petersburg Institute of Technology and had a Doctor of Mathematical Sciences degree from the Sorbonne. He made lasting contributions in the theory of elliptic and hyperbolic partial differential equations, potential theory, and problems of mathematical physics. His ideas were later applied by Henri Poincaré (1854–1912), a French mathematician, physicist, and philosopher, who had a major impact on the direction of 20th century mathematics. Henri Lebesgue (1875–1941), another influential French mathematician and one of the creators of modern measure theory, a man not usually given to praise, stated:

> Zaremba's scientific activity influenced so many research areas that his name cannot be unknown to anyone interested in mathematics. However, it seems that the power of the methods he created, and the originality of his imagination, can be appreciated best by those who work in the area of mathematical physics. There he showed his style and there his name is imprinted forever.

Given Zaremba's exalted status in academia and the lack of any written record, there have been heated debates on whether or not Banach, who was not a formal Jagiellonian University student, attended Zaremba's lectures. Two eminent Polish mathematicians who knew Banach well, Hugo Steinhaus and Kazimierz Kuratowski, felt that he did; and this opinion recurs in many other private communications. Andrzej Turowicz, who was very close to Banach, said:

> I always heard and believe that Banach attended Zaremba's lectures, but in my opinion this took place not before his departure for Lvov but after he returned to Cracow during the First World War. I was told, however, that Banach never attended any classes on a regular basis. This seems natural to me. He used to think with lightning speed, whereas Zaremba's lectures, with which I'm quite familiar, while scientifically superb, were pedantic and filled with little details. Banach had no need for long explanations; he grasped ideas instantaneously and was usually well ahead of his lecturers. On the other hand, from time to time he surely wanted to know what new topics might be introduced so that he could work on them himself.[8]

Banach spoke of Zaremba to Turowicz with great respect and admiration, and perhaps it was from these conversations that Turowicz got the idea that Banach audited Zaremba's lectures. In any case, Banach knew Zaremba's papers and main research results extremely well.

Further light on how Banach could have attended lectures at the Jagiellonian University without ever being officially noticed is shed by a differential geometer, Stanisław Gołąb, in an article about the university's history:

> ...We have to take into account the Jagiellonian University's curriculum in the forty years between 1895 and 1935. Before the introduction in 1926 of the so-called Master's Degree system, the study of mathematics was concluded either by a teacher's examination, in which case passing an exam in another subject (most often physics) was required, or by passing a doctoral examination. These two exams were independent of each other. Besides mandatory credits for recitation sessions, which appear sporadically in 1903–1904, and systematically in 1910–1911, and which were often just a formality, there were no other requirements for advancing from one year of studies to the next.
>
> Although so-called *colloquia* (midterm exams) existed, they were mandatory only for students seeking financial help. As a result of this system, professors essentially did not know their students. The graduates mainly entered the high school teaching profession. Academic and research careers, in view of the limited possibility for further study abroad, were accessible only to a select few who were financially independent, or to those who had enough stamina and determination to earn higher academic degrees (doctorate and habilitation) while performing teaching duties in secondary schools. A small number of mathematics departments at the universities in Cracow and Lvov and at the Lvov Polytechnic (there were no Polish universities outside the part of Poland under Austrian administration), provided some opportunities for an academic career.... While the lectures were not aimed exclusively at educating future high school teachers, by and large, there were rather few courses....
>
> One should also note that beginning with the academic year 1912–1913, there appeared a systematic course in mathematics for natural sciences (5 hours per week) offered by Włodzimierz Stożek. After that, mathematics courses for natural scientists entered permanently into the curriculum and were taught by W. Stożek, L. Chwistek, W. Wilkosz, and S. Gołąb.

During the first year of World War I the regular program of studies in mathematics was suspended. Besides two courses offered by Jan Śleszyński there were no other lectures. In 1915 the classes returned to normal and the schedule of courses again seems impressive.

Zaremba was still on the faculty, and Gołąb recalled that his teaching was characterized by absolute rigor and an insistence on an exposition of a subject's subtleties. His lecturing style employed long and convoluted sentences, whose logical progression became clear only after closer scrutiny. He enjoyed working on and solving difficult problems that bogged down other researchers. Always taking a philosophical view of a problem, Zaremba combined physical intuition with enormous erudition, a method that enabled him to connect seemingly unrelated problems. Whether or not Banach attended Zaremba's lectures, his own lecturing and working style were to be dramatically different.

$$* \quad * \quad *$$

The great intellectual events of those days included the appearance of Alfred Whitehead and Bertrand Russell's *Principia Mathematica*, which was published just before World War I, and Einstein's special and general theories of relativity. The Polish intellectual community was full of lively discourse. Inspiring, scholarly news flowed in from every direction.

Mathematics was alive with great moments. Young topologist Zygmunt Janiszewski (1888–1920), fresh from Paris, was formulating a program of action and goals for Polish mathematics. Kazimierz Kuratowski, the longtime director of the Institute of Mathematics of the Polish Academy of Sciences following World War II, describes this period in his *Half a Century of Polish Mathematics*:

> Towards the end of World War I, Kasa im. J. Mianowskiego, a savings and loan bank which served Polish scientists, especially those from the Russian-administered territories, founded the publication *Polish Science: Its Needs, Organization, and Development*. As the name suggested, its goal was to provide a forum for discussing the institutional organization of Polish science in a country that had only recently reestablished its independence. In the first volume of that publication, which appeared in 1918, Janiszewski published an article, "On the needs of mathematics in Poland," which with amazing clarity and precision

presented a blueprint for Polish mathematics. Janiszewski started out with the assumption that "Polish mathematicians do not have to be satisfied with the role of followers and customers of foreign mathematical centers," but "can achieve an independent position for Polish mathematics." One of the best ways of achieving this goal, suggested Janiszewski, was for groups of mathematicians to concentrate on the relatively narrow fields in which Polish mathematicians had common interests and – even more importantly – had already made internationally important contributions. These areas included set theory with topology, and the foundations of mathematics (including mathematical logic).

Janiszewski's youthful enthusiasm encouraged the realization of a very ambitiously formulated program:

> It is true that a mathematician does not require laboratories or any complex and expensive auxiliary devices. However, he does need an appropriate mathematical atmosphere and opportunities to stay in touch with his collaborators. Such an atmosphere can be created only if people work on common scientific problems. Collaborators are almost necessary for a researcher. Isolated, he most often withers. The reason for this is not just the psychological lack of incentives: an isolated scholar knows much less than those who work jointly. Only the final results reach him – ideas already mature and in their finished form, often many years after their creation, when they finally appear in print. An isolated person never realizes why they were created and where they came from; he did not live through the process of forging them together with their creators. We are far from those forges and boilers where mathematics is being created; we arrive late, and unavoidably we are being left behind. So the conclusion is that if we do not want to stay behind we have to grasp radical means and address the root of the problem. We need to create such a forge ourselves.[9]

In order to create an independent international position for Polish mathematics, Janiszewski analyzed in-depth interdependencies between different branches of modern mathematics, producing an original and intricate chart to visualize them. He further proposed the creation of a journal solely devoted to set theory and the foundations of mathematics. Such a journal, the first narrowly focused mathematical journal in the world, published in languages accessible to the international community, would fulfill a double role: it would present to the scientific world the achievements of Polish mathematicians, and at the same time it would attract papers by foreign authors with similar interests. In brief, it would become an international organ in the area of mathematics established by Poles. "If we want to achieve an appropriate position in world science, let's arrive through our own initiative," Janiszewski appealed to the Warsaw mathematical community. His goals were soon to be achieved. The first volume of *Fundamenta Mathematicae* appeared in 1920. Its editors were Stefan Mazurkiewicz and Wacław Sierpiński. Tragically, the volume also contained Zygmunt Janiszewski's obituary. He died on January 3, 1920, one of the many who fell victim to an epidemic of dysentery.

The journal was very well received in the mathematical world. Henri Lebesgue, discussing the foundations of Janiszewski's *Fundamenta Mathematicae*, advised Polish mathematicians to take into account "all possible applications of set theory, and not only direct applications that seem to follow from the programmatic proposals." This recommendation was motivated not only by his sincere desire for the new journal to succeed, but also by his concern about the future fate of set theory, one of the most important developing areas in mathematics at that period.

Lebesgue wrote that " . . . the theory was outside the domain of mathematics studied by the high priests of the theory of analytic functions," and that even if "that ostracism against set theory abates, this is only because set theory, which ironically grew out of the theory of analytic functions, turned out to be quite useful for her older sister and proved its value and its richness to all scholars." Also, thanks to contacts that the specialists in set theory maintained with those in other areas, "they did not go in areas of research isolated from the rest of mathematics. This could have led to the creation of a new, and perhaps a respectable discipline which might have remained outside the areas of general interest for a long time."

The positive effect of Lebesgue's advice, and his flexible interpretation of the limits of applications to set theory, was that many fundamental papers in diverse areas such as geometry, probability theory, and trigonometric series appeared in the pages of *Fundamenta Mathematicae*. In Lebesgue's words, set theory was repaying its debt to the theory of trigonometric series, since the *Cantor fractal discontinuum*, one of the most brilliant discoveries of Georg Cantor, was inspired by problems in the theory of trigonometric series.

In his memoirs Kuratowski, discussing Lebesgue's advice, comments that another positive effect was the liberation by 1929 of functional analysis (Banach's main domain of scientific pursuit) from *Fundamenta Mathematicae*. Research in functional analysis led to the creation of a special new journal – *Studia Mathematica*. But this is getting ahead of our story.

* * *

In this expectant, enthusiastic atmosphere, a serendipitous event occurred that proved to be a breakthrough in Stefan Banach's life. It happened in the spring of 1916 and involved his accidental meeting with Hugo Steinhaus. The meeting fundamentally changed Banach's professional and personal life; it was through Steinhaus that Banach was to meet his future wife, Łucja Braus.

Who was Hugo Dyonizy Steinhaus? He was born in 1887 in the Galician town of Jasło into a family of Jewish intelligentsia (his uncle was a well-known politician, member of the Austrian parliament). Although barely five years Banach's senior, his mathematical career was far ahead of the latter's. He obtained formal education in Lvov and then in Göttingen, where in 1911 he received a Ph.D. with David Hilbert for a thesis on trigonometric series. From 1920 until 1941 he was a Professor of Mathematics at Lvov University. Later, following World War II, he moved to the Western Silesian city of Wrocław, where he was a university professor and a member of the Polish Academy of Sciences. During this period he also visited several universities in the United States, including Notre Dame. Steinhaus, together with Banach, was to be the founder of the Lvov School of Mathematics. His bibliography contains over 170 articles with contributions in Fourier series, orthogonal expansions, linear operators, probability theory, game theory, and applications of mathematics to biology, medicine, electrical engineering, law, and statistics. He was also a well-known popularizer of science.

Kuratowski describes Steinhaus with admiration:

> Steinhaus was bestowed with an extremely valuable and rare abil-

ity to uncover mathematical elements in problems of natural sciences, economics, or engineering which he later formulated as mathematical theorems and conjectures. He did it with the uncontested mastery, finesse and brilliance, which so characterized him. At the same time he was able to interest numerous students and coauthors, creating a tradition of collaborative research in the area of applied mathematics.

Let us add that he was also able to explain his ideas with unusual clarity, not only to experts but to the population at large. His famous *Mathematical Miscellany* was translated into a dozen languages. The book, which in his opinion visualized mathematics, is a top achievement in the popularization of the subject.

But all of this happened later. At the time of his encounter with Banach, Steinhaus, after a brief military service in the Polish Legion that helped bring Poland its independence following the end of World War I, was preparing to take a position as Assistant at the [King] Jan Kazimierz University in Lvov. In his *Memoirs*, Steinhaus describes the encounter as follows:

> Although Cracow was still formally a fortress, it was already safe to promenade in the evening on Cracow's Planty. During one such walk I overheard [the] words "Lebesgue measure." I approached the park bench and introduced myself to two young apprentices of mathematics. They told me that they had another companion by the name of Witold Wilkosz, whom they extravagantly praised. The youngsters were Stefan Banach and Otto Nikodym.
>
> From then on we would meet on a regular basis, and given the presence in Cracow at that time of Władysław Ślebodziński, Leon Chwistek, Jan Kroć, and Włodzimierz Stożek, we decided to establish a mathematical society. As the initiator of the idea, I made my room available for meetings and, as the first step in preparations, nailed an oilcloth blackboard to the wall. When the French manager of the boarding house saw what I had done, she was terrified – what was the proprietor going to say? I calmed her down by reminding her that the owner of the building was my uncle's brother-in-law, and she forgave me my transgression. However, I had made a mistake. Mr. L. took the position of a traditional, hard-nosed landlord and was unmoved by the lofty goal the blackboard was supposed to serve. The society expanded – it was the first ray of light of this kind in Poland.

Hugo Steinhaus, throughout the rest of his life, in speeches and in writing, consistently claimed that Stefan Banach was the "greatest discovery" of his life. Shortly after their initial chance encounter, a very close collaboration developed between these two uncommon personalities and talents that was to last for many years. Their frequent meetings and discussions usually took place in the Esplanada Café on the corner of Karmelicka and Podwale Streets.

One afternoon Steinhaus described to Banach a problem that he had been trying to solve for quite some time. The problem concerned the convergence in the first moment of partial sums of the Fourier series of an integrable function. A few days later Banach returned with a complete solution. Steinhaus recalled:

> ...I was not a little astonished when, after a few days, Banach found a negative answer. The proof he gave me had a gap due to his unfamiliarity with the Du Bois-Reymond's example. Our joint note was presented by S. Zaremba to the Cracow Academy and, after considerable delay, it was published in 1918.

The paper appeared in the *Bulletin of the Cracow Academy*; it was Banach's debut as a mathematician.

The encounter with Steinhaus also had implications for Banach's personal life. On October 19, 1920 Banach married Miss Łucja Braus in a ceremony at the Cracow church of St. Szczepan-on-the-Sands. His bride came from a family of artisans and had started earning her living early. At the time she met Banach, she was a secretary for Władysław Steinhaus (Hugo's first paternal cousin), and later worked as a stenotypist at the law firm of Mr. Lisowski, Esq., the son-in-law of Ignacy Steinhaus, a lawyer and politician living in Vienna with whose relatives Łucja had been raised in Cracow. The first meeting between Stefan and Łucja took place in the apartment of Jadwiga Lisowska, where the Steinhauses lived at that time and where Łucja typed attorney Lisowski's legal arguments. Banach, who often visited there, found himself immediately attracted to young Łucja.

After the wedding the Banachs left for the resort of Zakopane in the southern Tatra Mountains, where they lived in the Villa Gerlach. The villa had been passed on from Dr. Chwistek, a physician and pioneer of tourism in the Tatra Mountains, to his son Leon Chwistek and to his sister Anna Stożek, wife of Włodzimierz Stożek, a mathematician who was on the faculty of the Jagiellonian University. In the summer the large mansion was occupied by the Chwisteks and the Stożeks, who in later years would be joined by the families of Stefan Banach and Wacław Sierpiński. In the nearby woods of sorb trees, on the banks of a cold mountain stream, Banach, Sierpiński, and Stożek wrote and proofread the mathematics textbooks on which whole generations of Polish schoolchildren and university students would be educated.

Banach was very much in love with his wife. She remained his closest and most loyal companion throughout trials and tribulations of the remaining twenty-five years of his life. He called her Lusia.

CHAPTER III

Lvov and Mathematics: 1919 – 1929

In the period from 1919 to 1929, Banach's phenomenal research and contribution to the reorganization of Poland's mathematical institutions earned him respect as one of Poland's leading mathematicians. He continued to meet regularly with Steinhaus, but he also met other mathematicians, formed friendships and established contacts in the elite mathematics community. With the end of World War I and the reestablishment of an independent Polish state, the mathematical community underwent profound changes in the way it collaborated on research. Banach was an important figure in this restructuring process.

The presentation of Banach's first paper at a session of the Cracow Academy of Sciences and its later publication in the *Bulletin of the Cracow Academy* (1918) drew the scientific world's attention to the appearance of a new and unquestionable talent. The paper carried considerable weight. Coauthored by Hugo Steinhaus, it was written in French, the *lingua franca* of science during that era, and was entitled "Sur la convergence en moyenne de séries de Fourier." It was a significant contribution to the then-fashionable theory of Fourier series, which was developed at the beginning of the 19th century to help solve such mathematical physics problems as heat transfer and wave propagation. The paper proved the existence of an integrable function whose Fourier (or harmonic) series does not converge in the mean. Today students of mathematics obtain this result as a relatively simple exercise, and we see it as a straightforward corollary to general functional analytic theorems later developed by Banach himself. The paper was perhaps one of the seeds that eventually bore fruit as that beautiful, abstract, and universal method that we now call functional analysis.

In the *Book of Minutes of the Mathematical Society of Cracow* we find a note, entered on April 2, 1919, to the effect that the Society's constitutional meeting convened at 5 P.M. in the quarters of the Philosophical Seminar at 12 Saint Ann Street. Stefan Banach was there as one of the founding members.

Franciszek Leja, later a professor of mathematics at the Jagiellonian University who occupied the chair of functions of complex variables and authored a calculus textbook familiar to generations of post–World War II Polish students, describes the first inaugural meeting, which was attended by 16 people:

> Professor S. Zaremba was elected chairman of the meeting and then Professor K. Żorawski submitted a resolution to establish a society under the name of the *Mathematical Society of Cracow*. . . . The goal of the Society was defined in the bylaws as follows: to stimulate research in pure and applied mathematics by holding meetings featuring lectures. . . . All those present at the constitutional meeting expressed interest in joining the Society and elected the first Board of Governors consisting of: S. Zaremba, president; A. Hoborski, vice-president; F. Leja, secretary, and I. Horodyński, treasurer. . . .

During 1919 the Society held eleven scientific (or ordinary) meetings and convened one Extraordinary General Assembly. Among the talks given at the ordinary meetings, two were delivered by Banach. On May 7 he spoke "On the theory of functions of a real variable," and on December 17 "On the theory of functions of the line." The lectures were followed by lively discussions.

During the first year of its existence the Society's membership increased from 16 to 50. The above-mentioned *Principia Mathematica* by Whitehead and Russell, and Einstein's general relativity theory continued to spark the scientific creativity of all natural scientists; but there is no information in the minutes indicating that special sessions of the Society were devoted to those topics, as was the case, for example, in Germany.

At the Extraordinary General Assembly on November 29, 1919, President Zaremba read a letter from Lvov mathematician Antoni Łomnicki. It stated that the Lvov Mathematical Society had been established two years earlier, and that its members would like to join the Cracow Society if the latter were interested in extending its activities to all of Poland. Samuel Dickstein, a senior member of the Society, then reminded members that as early as 1880 a Circle of Polish Mathematicians had been founded in St. Petersburg, and that it had published four volumes of scientific papers authored by its members. That publishing activity was later taken over by the journal *Prace Matematyczno-Fizyczne*[1] which was established in 1888 in Warsaw.

On December 22, 1920 new bylaws acknowledged that out-of-town members could form chapters of the Society, which had already been renamed the *Polish Mathematical Society*. One of the new chapters was organized in Lvov.

* * *

At that time, thanks to Steinhaus' mentoring and friendship, other luminaries of Polish mathematics became familiar with and curious about Banach's work. After all, according to Steinhaus that self-made man was solving difficult and convoluted mathematical problems on a regular basis.

During this period, Banach wrote his second paper, this time on his own: "Sur la valeur moyenne des fonctions orthogonales." In this one he proved that the sequence of arithmetic averages of an orthonormal sequence of functions is almost everywhere convergent to zero – a phenomenon similar to the strong law of large numbers in probability theory. Later this result was considerably strengthened by other mathematicians.

It was also observed that an application of the Kronecker lemma significantly simplifies the proof so that Banach's paper subsequently lost its
independent importance.

In the same year Banach wrote another paper, its elegance still appre-
ciated to this day. It appeared later in *Fundamenta Mathematicae* under
the title "Sur l'équation fonctionelle $f(x + y) = f(x) + f(y)$." The
problem of finding functions satisfying the above functional equation
had been the subject of investigation by many mathematicians, beginning with the 19th century effort of Augustin Louis Cauchy (1789–1857).
Cauchy was one of the first mathematicians to place Newtonian calculus on a rigorous foundation. He is also infamous among present-day
calculus students as the creator of the feared $\epsilon - \delta$ definition of a continuous function. Banach's two-page note demonstrated that measurable
solutions of the above equations are necessarily continuous and, consequently, have to be linear functions. This result is still used and continues
to find applications in several areas of mathematical analysis. His next
paper, "Sur les ensembles de points ou la dérivée est infinie," contains
a proof of the theorem that asserts that the set of points where the right
derivative of an arbitrary function is infinite is of measure zero. This
result was much stronger than the previously known theorems which
assumed that the function was continuous and where only two-sided
derivatives were considered. At the same time, Banach's proof of that
stronger result was simpler than the previously known proofs of weaker
results.

The last two papers were subtle masterpieces of the modern theory
of functions of real variables, and began the whole series of Banach's
articles devoted to that field. These discoveries alone would have assured
Banach the respect of other mathematicians.

In 1920 Banach was offered an assistantship at the Lvov Polytechnic by Antoni Marian Łomnicki (1881–1941), who himself had recently
been appointed Professor of Mathematics there. Łomnicki also wrote
many textbooks and papers in the area of geometry and cartography.
This was Banach's first paid job in academia, although his duties included baby-sitting for Łomnicki's infant daughter and perhaps other
unorthodox chores.

Banach continued working at a frantic pace. He befriended Stanisław Ruziewicz, and in their joint paper, "Sur les solutions d'une équation fonctionelle de J. Cl. Maxwell," they found all functions f that satisfy the equality $f(u)f(v)f(w) = u^2+v^2+w^2$ for all numbers u, v, w. The equation had been introduced by Maxwell in connection with problems of mathematical physics, and he had solved it for the special case of differentiable functions. Ruziewicz and Banach's solution was not only more general but also relatively simple and elementary.

Banach's sixth paper, "Sur les fonctions dérivées des fonctions mesurables", returned to the theory of functions of a real variable. The paper brought a new impetus to the study of derivatives of very general functions, and the old classical theory of differentiation was merged with the new Lebesgue theory of measure and measurable functions. This pattern would become the hallmark of many Banach contributions; boldly and ingeniously he would apply newly emerging disciplines to solve old classical problems. Both the old and the new theories would benefit from these developments.

* * *

During part of this period of intense research activity, Banach was living with and working for his professor. Łomnicki's older daughter Irena notes:

> I was then a young lady in my teens when, in 1920, Banach arrived in Lvov and moved into a room in our house at 19 Nabielaka Street. In addition to a kitchen and a bathroom, our apartment had three rooms: a dining room, a bedroom, and my father's study. As far as I can remember, Banach slept in my father's study. He did not live with us for very long – perhaps for a few months – only until he set himself up independently in Lvov. He dressed very modestly in a patched-up suit and shoes. He gave the impression of a pleasant gentleman who would not disturb other household members. My mother Władysława loved literature and owned a lot of fiction. Sometimes Banach would borrow books from her, and my mother often expressed surprise at how fast he would finish them and how many vivid details he would retain from his readings.

Much later, when Banach's international reputation had already been firmly established, Antoni Łomnicki liked to tell stories about the phenomenal memory of his former assistant who could reconstruct in detail the contents of novels read several years earlier, or quote from memory whole fragments of those books – and all of this done with Banach's characteristic slightly amused, skeptical smirk.

Also in 1920, after long hesitation, Banach finally wrote his doctoral dissertation entitled "On operations on abstract sets and their applications to integral equations." It was written in Polish and in 1922 its French version was published in *Fundamenta Mathematica* as "Sur les opérations dans les ensembles abstraits et leur application aux équations integrales."

This work deserves a thorough description not only because it was Banach's Ph.D. thesis, but also because it contained some of the most important ideas he ever produced. So what is in it? First of all, it introduces an abstract object that later came to be called a *Banach space*. Banach gave an axiomatic definition of such possibly infinite dimensional spaces and introduced the notion of a linear transformation of them. A variety of concrete examples encompassing situations that had previously been considered independently of each other now fell under the unifying umbrella of Banach spaces. The familiar Euclidean, finite-dimensional spaces were obvious special cases, but their structure was relatively simple, and the only linear transformations of them were translations, rotations, and reflections. Much more exciting were the infinite-dimensional Banach spaces of functions important in many areas of classical analysis. Their structure, and the structure of linear transformations (operators) on them, was much richer and is not fully understood even today. Although it was Banach's first paper in the area that is now called *functional analysis*, his axiomatization was amazingly complete and, in a sense, final. It should be said that, to some degree, his dissertation brought functional analysis to independent life in a single sweep.

It is true that similar ideas of axiomatization were maturing in papers by other mathematicians as well. In 1921, the Austrian mathematician E. Helly introduced similar concepts. Steinhaus himself introduced the notion of an abstract linear operator, and Norbert Wiener, in his autobiography, also lays a partial claim to discoveries in this area. What distinguished Banach's work from all the other papers on similar topics was his explicit demand for completeness of these spaces. He recognized it as their fundamental property and demonstrated why that property is essential for obtaining deep and useful theorems. Two such seminal theorems appeared in his dissertation. The first asserted that the point-wise limit of a sequence of linear and continuous operations is necessarily linear and continuous as well. The second was what is now known to every mathematician as the *Banach fixed point theorem* on contracting operations. This theorem, an abstract and very general version of the classical method of successive approximations that had been used earlier in many concrete situations, gives a practical method of finding solutions to equations of various types. Both of these theorems are like master keys that fit many doors without which each door would have had to be opened separately by a locksmith. Banach's doctoral dissertation was a cornerstone on which he, his students, and others erected the edifice of functional analysis in the decades to follow.

Despite the importance of Banach's discovery, the formal process of obtaining a Ph.D. was another matter. There were obstacles. Banach, who was a self-taught man and had no college degree, had to obtain a special legal exemption from the Minister of Education to be admitted directly to his master's degree examination, which had to precede his doctoral examination. But that was not the biggest impediment. Otto Nikodym who after World War II became a faculty member at Kenyon College in Ohio, told Andrzej Turowicz that Lvov professors realized that the material for Banach's dissertation had been ready for some time, but that Banach had no intention of writing it down. Once Banach had proved his theorem, he was not very interested in turning it into a publishable paper. The process bored him. He was fascinated by mental speculations but abhorred the chores of putting them down neatly on paper. Nikodym recalls how Banach's colleagues helped, or tricked, him into completing a publishable paper. This was recorded by Turowicz:

> Professor Ruziewicz instructed one of his assistants to accompany Banach on his frequent visits to the coffee houses, query him in a discrete fashion on his work, and afterwards write down Banach's theorems and proofs. When all of this information was typed out, the notes were

presented to Banach, who then edited the text. This is how his Ph.D. dissertation was finally completed.

At the age of 28 Banach became a doctor of mathematical sciences. His advisor was Antoni Łomnicki.

Stanisław Marcin Ulam, a Lvov mathematics student who later made a significant contribution to the Manhattan Project and taught at Harvard, Columbia, and the University of Colorado at Boulder, describes in his memoirs how Ph.D. degrees were awarded in those days at the Jan Kazimierz University:

> The ceremony was a rather formal affair. It took place in a large institute hall with family and friends attending. I had to wear a white tie and gloves. My sponsors Stożek and Kuratowski each gave a little speech describing my work and the papers I had written. After a few words about the thesis, they handed me a parchment document.

A doctoral examination in Lvov covered both the candidate's main field and also an additional topic. Steinhaus talked Banach into selecting astronomy as the extra subject. Banach, initially full of good intentions, reported to Professor of Astronomy Stanisław Loria to ask for the exam's syllabus. Loria gave him a list of readings and later professed to have been very impressed by Banach's astounding daily progress in mastering the assigned material. But although Banach enthusiastically and successfully went through the learning process, he did not think much of the idea of taking an exam on the subject, and subterfuge had to be used to drag him to the examination.

*   *   *

After earning his doctorate, Banach entered a period of even more intense work. At the same time his earlier results began to appear in print. His articles published in *Fundamenta Mathematicae* and the *Bulletin International de l'Académie Polonaise des Sciences et des Lettres* were being distributed abroad, and naturally helped expand his foreign contacts. An English language paper entitled "An example of an orthogonal development whose sum is everywhere different from the developed function" appeared in 1923. The result in that paper is surprising in its simplicity. It says that for each integrable function f that is not square integrable, one can construct a complete orthogonal system consisting of bounded functions such that all the Fourier coefficients of the function f with respect to that system are equal to zero. Later that theorem became a starting point for several papers by other authors.

In the Lvov Polytechnic bulletin for the academic year 1920–21, Stefan Banach is mentioned as an assistant in the Department of Mathematics headed by A. Łomnicki. Banach's teaching obligations of three hours per week did not slow down his research. The free exchange of ideas and thoughts so characteristic of the Lvov mathematical community from the very beginning bore fruit quickly. In that period he wrote a paper, "Sur le problème de la mesure," which was presented on February 27, 1922 at the scholarly session of the Faculty of Mathematics and Natural Sciences, and later published in the fourth volume of *Fundamenta Mathematicae*.

The problem of measuring surface areas of solids and planar figures goes back to antiquity. From the dawn of history, people somehow managed to measure the surface area of simple geometric objects such as rectangles, triangles, or even arbitrary polygons. Archimedes' major achievement was the invention of a method of successive approximation to compute the surface area and volume of more complicated objects such as circles, balls, and cones. However, his ideas were not rigorously proven until the 19th century when Camille Jordan significantly extended the set of figures that could be measured. That set was further widened by Henri Lebesgue at the beginning of the 20th century. He showed that the surface area in a plane, and ultimately the volume in a three-dimensional space, can be assigned in a very satisfactory way to practically any object that can be effectively described (or "constructed," as mathematicians like to say). An unsolved problem that puzzled Lebesgue for quite some time was: Is it possible to assign an area to any set in the plane, and assign a volume to any set in the three-dimensional space? The only demand was that the measurement system satisfy two very natural requirements: that the areas (or volume) of congruent objects are the same, and that the area (volume) of the union of disjoint sets is equal to the sum of the areas (volumes) of component sets. In 1914 Felix Hausdorff had shown that this cannot be done for volume. Employing a very ingenious method, Banach demonstrated that it also cannot be done in the plane, or even for an interval. This result not only solved the problem but, as in other instances, became a starting point for a series of papers by other mathematicians. It was in this paper that two now commonly known notions were born: the *Banach integral* of an arbitrary bounded function and the *Banach generalized limit*. Research in these areas continues to this day. However, Banach's measurement method, although important, did not play as critical a role in mathematics as the Lebesgue measure. The reason became apparent very soon. If one wants measures to preserve good properties when subject to certain infinite limit operations, then the class of measurable sets introduced by Lebesgue turns out to be the broadest possible.

*　*　*

In 1922 Stefan Banach received his habilitation. The official *Chronicle of the Jan Kazimierz University* in Lvov for academic year 1921–22 reported in "Personnel Items":

> On April 7, 1922, by resolution of the Faculty Council, Dr. Stefan Banach received his habilitation for a Docent of Mathematics degree. He was appointed Professor Extraordinarius of that subject by a decree of the Head of State issued on July 22, 1922.

So, almost immediately after his habilitation, Banach became a Professor Extraordinarius at Jan Kazimierz University.[2] Through his appointment at the age of 30 as a professor at a dynamic and well-respected university, Banach bettered his social standing faster than he could ever have expected. His financial, intellectual, and social situations were now firm. Despite his young age, he was a distinguished and well-known scientist, well liked in Lvov's scholarly circles.

By and large university professors lived very comfortably. The cost of living in Lvov was less than in Warsaw, and materials provided at the time to visitors by the Scientific Society in Lvov indicate that a person could live on about one dollar per day, which included rent for a comfortable apartment. Persons with more modest demands could live reasonably on half a dollar a day.

As a university professor Banach should have been in good financial shape. However, such was not the case. He started living beyond his means and was often in debt. To improve his financial position he took to writing textbooks, a task that he took very seriously. Steinhaus comments on this "textbook" period in Banach's life:

> Most of Banach's time and energy was consumed by writing arithmetic, algebra, and geometry textbooks for high schools. He was writing them with Sierpiński and Stożek, but also by himself. Banach, because of his experience as a high school tutor, realized perfectly well that each definition, each proof, and each exercise is a major challenge for an author of a school textbook who cares about teaching.

Later on, Steinhaus makes a curious but also telling comment about Banach's textbook activity: "In my opinion, Banach lacked only one of many talents needed by an author of school textbooks: an aptitude for spatial imagination."

In the years 1929 and 1930 Banach also published two volumes of an academic textbook, *Differential and Integral Calculus*. It was well known for its rigorous approach and filled an acute gap in the Polish mathematical textbook literature.

*   *   *

Despite these "side" activities, Banach continued an active research program that generated four more papers.

In "Sur un théorème de Vitali" he gives a simple proof of an important theorem of measure theory and simultaneously solves a problem posed by another well-known mathematician. Ideas from this paper were to be utilized for years to come. The paper, "Sur une classe de fonctions d'ensemble," contains a significant contribution to the theory of the differentiation of general functions. "Un théorème sur les transformations biunivoques" was a contribution to an important theorem in set theory, originally due to Schröder and Bernstein.

The last paper, "Sur la décomposition des ensembles de points en parties respectivement congruentes," merits closer attention. Banach wrote it with Alfred Tarski, a philosopher and mathematician 10 years younger than himself (who later became Docent of Warsaw University and in 1946 became a professor at the University of California at Berkeley; Tarski was a founder of a research school in mathematical logic and the foundations of mathematics that has had a decisive influence on the development of these subjects in the United States and elsewhere). The paper contains an intuitively shocking result that sheds new light not only on the theory of measuring sets but also on our understanding of many facts at the very foundations of mathematics.

To explain this result it helps to introduce two definitions. We shall say that two subsets of a ball are congruent if one can be transformed into another by rotating the ball about its center. We shall also say that two sets are equivalent by a finite decomposition if both of them can be split into the same number of parts in such a way that the respective parts of these two decompositions are congruent.

Banach and Tarski demonstrated the existence of two disjoint subsets of the ball, each of which is equivalent to the whole ball by a finite decomposition. A less precise but more descriptive explanation of the result can be phrased as follows: The ball can be split into pieces in such a way that from these pieces one can assemble two balls exactly the same as the original ball. It follows, obviously, that for these subsets the notion of volume does not make any sense – if it were otherwise, the volume of the original ball would be twice the volume of each reassembled ball.

This paradoxical decomposition of the ball, now called the *Banach-Tarski paradox*, was one of the discoveries that reinforced the need for new research into understanding the fundamental notions in mathematics. In particular, it focused attention on the role of the so-called *axiom of choice*. It became clear that in the above paradoxical decomposition of the ball, one guarantees its existence but does not show how to construct it. Today we know that its construction is impossible. Not only did the paradox inspire mathematical research around the world, but the result also generated interest beyond the professional mathematician's. It was one of the few nontrivial modern mathematical discoveries that could be easily described to lay people.

These years were very happy ones in Banach's life. He was honored, elected to membership of scientific societies, and invited to deliver lectures. In the directory of the Transportation Faculty at the Lvov Polytechnic we find Banach listed as a lecturer on general and theoretical mechanics. In this capacity he taught three hours and also conducted two hours of recitation sessions per week. In 1924 he was elected a corresponding member to the Academy of Knowledge (which 15 years later awarded him its Grand Prix). Shortly thereafter he left to give a series of lectures in France.

A report on the French trip appears in the 1924 *Chronicle of the Jan Kazimierz University*:

> This year, as in previous years, the Ministry of Higher and Public Education granted leave to a number of professors of the Faculty of Mathematics and Natural Sciences so that they can familiarize themselves firsthand with the latest discoveries in science and establish personal contacts with its representatives in other countries. For the full 1924–25 academic year, sabbatical leaves (with full salary) were granted to Professor Loria to continue his research at the California Institute of Technology in Pasadena, and to Professor Banach to continue his mathematical research in Paris.

The same *Chronicle* provides further information on Banach. First, it mentions the split of the Faculty of Philosophy into the *Facultas Literarum* – Faculty of Humanities – and the *Facultas Scientiarum* – Faculty of Mathematics and Natural Sciences. It also reports that at the October 1, 1924 meeting of the Faculty Council, Professors Żyliński, Steinhaus, Ruziewicz, and Banach were appointed to chairs in the Faculty of Mathematics and Natural Sciences. Despite his likely absence from campus at that time, the Council, at its next meeting on November 12, 1924, also passed a unanimous resolution recommending that Stefan Banach be promoted from Professor Extraordinarius of Mathematics to Professor Ordinarius. At the age of 32 Banach had achieved the highest possible rank in the scientific community, although to tell the truth, ranks, positions, titles, and honors were of little interest to him.

We do not really know how important Banach's trip to France was for his further creative development. Marek Kac (1914–1983), a student of Hugo Steinhaus, who became a distinguished probabilist and mathematical physicist at Cornell, Rockefeller, and Southern California Universities and a member of the United States National Academy of

Sciences, described the influence of French mathematics on the Lvov school:

> The influence of Lebesgue on the Lvov school was very direct. The school, founded a little later than the Warsaw School by Steinhaus and Banach, concentrated mainly on functional analysis and its diverse applications, the general theory of orthogonal series, and probability theory. There is no doubt that none of these theories would have achieved today's level of prominence without an essential understanding of the Lebesgue measure and integral. On the other hand, the ideas of the Lebesgue measure and integral found their most striking and fruitful applications there in Lvov.

During this period, although Banach's discoveries were moving well beyond the traditional scope of French mathematicians, the French episode in Banach's life is well worth a more detailed investigation.

*   *   *

We learn from the *Chronicle* that during the 1925–26 academic year Banach taught at the Jan Kazimierz University a course on infinitesimal calculus and directed recitation sessions in the same subject. He was also the faculty advisor for out-of-town members of the Mathematical-Physical Circles of the Polish Academic Youth Organization.

Banach remained an active member of the Lvov Branch of the Polish Mathematical Society. On March 3, 1926 a special session was devoted to his publications, more and more of which were appearing in print:

"Sur les lignes rectifiables et les surfaces dont l'aire est finie" is an important paper on curves of finite length or, more precisely, on functions of finite variation, and on some of their generalizations for functions of two variables. In this paper Banach presented his elegant formula for the variation of a continuous function, which shows that the variation is equal to the integral of the indicatrix of that function (the indicatrix of function f is a function that at each point a is equal to the cardinality of the inverse image of point a via function f). That formula found an application in, among other areas, the theory of random measures.

The paper "Sur une propriété caractéristique des fonctions orthogonales" contains a striking application of a theorem from Banach's doctoral dissertation on limits of sequences of linear operators to the theory of orthogonal series.

"Sur le prolongement de certaines fonctionelles" solves a problem on the extension of certain functionals posed by the French mathematician Paul Lévy. The problem was perhaps not very important in and of itself, but Banach was interested in the application of his modern tools to its solution.

"Sur la convergence presque partout de fonctionelles linéaires" responds to a concrete problem about a sequence of integral transformations posed by Lebesgue, and Banach again used his extremely fruitful functional analytic approach. His result deals with the limits of linear operators acting from a Banach space into the space of all measurable functions.

"Sur une classe de fonctions continues" contains a study of continuous functions in the measure-theoretic context and a partial solution to a problem posed by the Russian mathematician Luzin.

The next paper, "Sur le principe de la condensation de singularités," written jointly with Steinhaus, was published in 1927 in *Fundamenta Mathematicae* and remains one of Banach's best known works. In it Banach and Steinhaus prove a generalization of Banach's doctoral dissertation theorem on a sequence of linear operators. At the same time, with the help of Stanisław Saks, the proof was significantly simplified and became short and elegant. Reviewing the paper, Saks noted that the authors had repeated Baire's argument to the effect that the n-dimensional Euclidean space is not a countable union of nowhere dense sets or, in other words, it is not a set of Baire first category. Because that latter fact is also valid for Banach spaces, it permits an almost instantaneous proof of the theorem for the sequence of linear operations. The method of categories turned out to be extremely useful in functional analysis. Banach used it later many times and "squeezed all the juice out of it." Today the theorem on sequences of linear operations is simply called the Banach-Steinhaus theorem, and is considered one of three fundamental theorems of functional analysis.

The same paper also contains an abstract method called the principle of condensation of singularities, which permits an easy proof of the existence of many counterexamples in analysis – for example, the continuous function for which the Fourier series diverges on a dense set.

Besides dutifully working the Society's lecture circuit, as reported in *Annales de la Sociéte Polonaise de Mathématique*, Banach continued to be responsible for a number of teaching duties. According to the 1926–27 *Chronicle*, the Faculty of Mathematics and Natural Sciences featured four active mathematical seminars: Seminar A headed by Professor Żyliński; Seminar B by Steinhaus; Seminar C by Ruziewicz, and Seminar D by Banach. The same *Chronicle* notes that Banach was the faculty person responsible for leading the young people's Mathematical-Physical Circle and that Steinhaus was in charge of the Society of Jewish Students of Philosophy. The *Chronicle* also describes Banach's seminar:

> At the seminar, original works were presented, mainly in the areas of functions of a real variable, the theory of functional operations, set theory, and topology. There were 45 participants. During the recitation sessions students solved problems assigned during lectures at the blackboard. During the last year, the seminar purchased 13 books (17 volumes) for a total sum of 310 złotys and 80 grosz. In addition, 16 reprints of papers were received as gifts from the authors.

The 1928–29 issue of the *Chronicle* notes that Stefan Banach, on behalf of the University, acted as chairman of the joint University and Polytechnic Board for Popular Lectures, and that members of that board were also listed as members of the Faculty Senate. In the reports of the department heads of the Faculty of Mathematics and Natural Sciences we read, in the stilted, official prose:

> Mathematical Seminar D Head: Professor Dr. Stefan Banach. Auxiliary scientific personnel: Zenon Wojakowski, demonstrator. A mathematical seminar devoted to functions of a real variable was conducted in which participants presented specialized papers from the area. Also, recitation sessions on general mechanics were held where students solved problem at the blackboard and had to turn in written assignments that had been given out at a parallel lecture on the same subject. The number of participants: (a) at the seminar – ca. 10, (b) at the general mechanics recitation sessions – ca. 200.

The memoirs of Marceli Stark, who following World War II became one of the chief organizers of the flourishing Polish mathematical publishing enterprise and must have coaxed scores of mathematicians into becoming authors of monographs and textbooks, shed light on how mathematical instruction was carried out at the Jan Kazimierz University. Here is a perhaps long but interesting quotation from his memoirs, which were published in *Wiadomości Matematyczne*:

> At the Jan Kazimierz University there were at the time four chairs in mathematics. They were held by E. Żyliński, H. Steinhaus, S. Ruziewicz, and S. Banach (given here in order of appointment). Customarily the freshman mathematical analysis course was taught each year by a different professor who then continued the analysis course for the same class during the sophomore year, and also conducted seminars and supervised the same students throughout their third year. After that, during their senior year, students were supposed to attend only special topics courses and seminars.
>
> My class was guided by Professor Steinhaus. It was a very big class, and the analysis lecture was attended by over 220 students squeezed into a smallish and poorly ventilated lecture room, standing in the aisles, and sitting on the window sills. The whole mathematics department had only four teaching assistants (one senior, one junior, and two so-called auxiliary), and as a result the class had not been split into smaller groups for the recitation sessions, which were conducted by Professor Steinhaus personally, without any help from assistants.

Describing Steinhaus' lectures Stark wrote:

> His figure, perched high on the podium by a small five by five foot blackboard dominated the crowded room. . . . Despite Steinhaus' attention to preparation, the lectures were too difficult for the average student. In this respect Steinhaus was the opposite of Banach, whose lectures were not only meticulously prepared, but also delivered in such a way that there was probably no one in the classroom who could not follow them.

Later in the memoirs Stark describes the personal relations among Lvov's university professors:

> I would like to say a few words about Professor Steinhaus' relationship with the other mathematicians of the Lvov mathematical school.

There exist certain misunderstandings, especially concerning the relationship between Banach and Steinhaus. . . . It has been said that Banach turned to functional analysis under Professor Steinhaus' influence. I was curious about this for a long time and, shortly after World War II, asked Professor Steinhaus if that indeed was the case. Steinhaus contemplated the question silently for a few minutes and then curtly said "No!" Banach went in that direction independently, as he did when he started working on set theory and, earlier, on the Lebesgue integral (he was still a student then, and the Lebesgue integral was not yet included in the textbooks). Steinhaus unquestionably exerted a major influence on Banach: Banach's first paper on trigonometric series was the outcome of a problem posed to Banach by Steinhaus; under Steinhaus' influence Banach started working on the orthogonal series. At the beginning of the twenties they worked with each other very closely. In time, as the structure of functional analysis became larger and larger, their paths started diverging. Banach assembled a group of his own disciples such as Schauder, Mazur, Orlicz (the latter was also Steinhaus' student), and Ulam (who started out under Professor Kuratowski's direction), who worked in functional analysis and its applications.

Between September 7 and 10, 1927 the First Polish Mathematical Congress was held in Lvov. Stanisław Warhaftman described the Proceedings in *Mathesis Polska*:

> The Congress assembled over one hundred people from the faculties of universities and gymnasia. Besides Polish scientists, foreign scholars participated. These included Aksel Andersen (Copenhagen), Nina Bari (Moscow), Vaclaw Hlavaty (Prague), Leon Lichtenstein (Leipzig), Nikolai Luzin (Moscow), Dimitrii Menshov (Moscow), Moses Jacob (Vienna), John von Neumann (Budapest), and Pierre Segrescu (Bucharest).
>
> The activities of the Congress were conducted in the following sections:
>
> 1) mathematical logic and the foundations of mathematics,
>
> 2) algebra and number theory,
>
> 3) theory of functions of real variable,
>
> 4) analysis,
>
> 5) geometry,
>
> 6) applied mathematics,

7) mechanics and mathematical physics,

8) astronomy,

9) didactics, history, and philosophy of mathematics, and

10) general interest section.

The largest number of presentations were given in the sessions on mathematical logic and the foundations of mathematics, set theory and functions of a real variable, and analysis. Taking into account the fact that the number of talks in the first two of the above-mentioned sessions actually made up 40% of the total number of lectures, one can conclude that the Congress was dominated by the research areas cultivated at Warsaw University, which comprised problems in set theory, topology, logic, and related topics. . . .

In the sessions on set theory and functions of real variables, the Warsaw and Lvov contingents dominated. In a sense they could be grouped as one large school representing the general direction of Polish mathematics.

> Polish mathematician Wacław Sierpiński was chairman of the Warsaw delegation and gave a paper entitled "Remarks on the Egorov theorem." In addition, lectures and talks were given by several docents of Warsaw University: Kazimierz Kuratowski, Bronisław Knaster, Stanisław Saks, Stefan Straszewski, Alfred Tarski, and Antoni Zygmund. The Lvov group was represented by Professor Stefan Banach, who delivered a lecture "On the method of Cauchy majorants in the theory of functionals," and two other talks, "On conditionally convergent series of functions" and "A short proof of existence of the characteristic value of a symmetric kernel." Professor Ruziewicz gave a talk "On functions satisfying the generalized Lipschitz condition," Professor Hugo Steinhaus spoke on "An application of functional operations to the problems of the theory of orthogonal series," and Professor Włodzimierz Stożek spoke "On fixed points under continuous mappings." From the younger generation, lectures by Dr. Stefan Kaczmarz entitled "Conditions for convergence of orthogonal series" and by Stanisław Mazur "On the summability methods" should be mentioned.

In addition, in the section on didactics of mathematics, Banach delivered an address "On the notion of limit." Clearly, he was one of the most active participants among the assembled mathematical luminaries. The article in *Mathesis Polska* continues to say:

As far as general impressions are concerned, the Congress demonstrated clearly that set theory and research on the foundations of mathematics are "specialties" of Polish mathematical thought. The appearance of the tenth volume of *Fundamenta Mathematicae* emphasizes this point further.

The author also noted with regret that

...the beautiful area of applications of mathematics, including above all mechanics, but also the whole discipline of mathematical physics, has in Poland, besides rare exceptions, very little understanding, in particular among the younger generation of mathematicians.

In November of 1927 Banach published a communication "Sur les équations à infinité d'inconnues I" and in 1928–29 Banach and Steinhaus jointly taught a course in "Selected topics in mathematical physics;" Banach by himself taught analytical mechanics, theory of functionals, and an advanced seminar. A year later he taught advanced analysis (including recitation sessions), general mechanics, and the dynamics of rigid bodies, and he conducted the lower mathematical seminar.

The preceding pages indicate that the collaboration between Banach and Steinhaus had not ended entirely, and in some way had even become deeper. One result of this cooperation was their joint founding and publication, in 1929, of what later proved to be an influential journal, *Studia Mathematica.*

CHAPTER IV

Studia Mathematica and *Opérations Linéaires*

THOUGH BANACH CONTINUED TO PUBLISH numerous research papers during the decade before World War II and to participate in the leadership of the Polish mathematical community, his two most important contributions to the mathematics of this period were the creation, together with Hugo Steinhaus, of the journal *Studia Mathematica* and the publication of his book *Théorie des Opérations Linéaires*. *Studia* was supported initially by a grant from the Ministry of Religious Denominations and Public Education and, in the last year before World War II, by the Józef Piłsudski Fund for National Culture. Volume IX appeared in 1940 already under the auspices of Soviet authorities. Kazimierz Szałajko, Banach's assistant beginning in 1935 (and currently a docent emeritus at the Silesian Polytechnic in Gliwice, Poland) who was responsible for the adminitration of *Studia Mathematica*'s assets from 1934 until 1939, reports that the financial income from subscriptions was substantial and the balance of the journal's account maintained at Bank Handlowy grew rapidly. There was also a very active exchange program with other mathematical journals, which directly benefited the departmental library.

The September/October 1929 issue of *Mathesis Polska* contained information about the creation in Lvov of a new mathematical publication:

> The first volume of a new periodical, *Studia Mathematica*, appeared in Lvov under the editorship of Stefan Banach and Hugo Steinhaus. The journal will publish original papers in the areas of pure and applied mathematics, and the editorial board plans to focus on research in functional analysis and related topics.
>
> *Studia Mathematica* publishes papers only in French, German, English, and Italian. Papers should be addressed to one of the editors: Professor Stefan Banach, Lvov, 4 St. Nicolaus Street, or Professor Hugo Steinhaus, Lvov, 14 Kadecka Street. Price of one volume is $1.50 abroad and 12 złotys in Poland. The main office for the journal is located in Lvov, at 4 St. Nicolaus Street, The University.

Note that Banach gave his office address whereas Steinhaus gave his home address; the Banachs had an apartment in the "old" university building. There was a request that corrected galley proofs be returned to Steinhaus, who took care of all the minute details of the publication process: contacts with the printing houses, selection of fonts, finances, exchange programs with foreign journals, and so on.[1]

Here it seems appropriate to comment on a certain style that characterized Hugo Steinhaus' relations with other people, a style that went beyond the transmission of facts and added a deeper dimension to his personal interactions.

The following quotation from the memoirs of Andrzej Turowicz illustrates a side of Steinhaus' personality that synthesized pragmatism and a special sense of metaphysics, two characteristics that may be necessary for a mathematician to go beyond a conventional career:

> Let me quote here an anecdote which I heard before the war from Professor Steinhaus at the Scottish Café in Lvov and which was also repeated in his memoirs. Professor Steinhaus asked me, "Do you know what is *the* most important question of all times?"
>
> I answered that I had no idea, and he then told me that once, during a lecture, Hilbert had said that if, touched by a magic wand, he fell asleep and woke up 500 years later, he would not ask about historical events

or social changes, but would inquire about what had been learned about the distribution of zeros for the Riemann function, because this was *the* most important problem . . . period.

Professor Steinhaus had, although what I'm going to say may sound pretentious, a method similar to Socrates'. He liked to lead people to a deeper understanding of a problem by asking them questions. And he would pin them to the wall with those questions so that they often ended up speechless. In his Göttingen reminiscences I believe he recalled an encounter with his landlady, whom he asked why one had to pay taxes. She answered:

– Ahemm, one pays taxes to maintain the standing army.

– But what do we need this army for?

– Well, I guess to beat the French.

– But what do we need to beat the French for?

And, at that point, she ran out of answers. He used that method very often.

Professor Steinhaus liked to teach young people to work neatly and accurately. I will repeat what I heard from Professor Mazur who, while still a student, wrote his first mathematical paper and submitted it to Professor Steinhaus. The latter was supposed to present it at a meeting of the Lvov Scientific Society, which was to convene that evening. A couple of hours before the meeting Professor Steinhaus summoned Mazur and said:

– Mr. Mazur, how can I present your paper if you failed to give it to me?

Mazur was flabbergasted.

– What do you mean, Sir, I personally handed you the paper.

– No, you handed me four sheets of blank paper.

By now, Mazur was completely dumbfounded. Professor Steinhaus pressed him,

– Well, take a look.

Here is what happened. Before the war one could buy a cheap kind of paper, called "concept" paper, which had a slightly yellowish tint. Mazur had run out of ink while writing the paper, and since he did not feel like running to a drugstore to get some more, he simply added water to the inkstand.

Then Professor Steinhaus took Mazur's paper and turned to the poor author.

– Well, Mr. Mazur, perhaps there is something written here after all. But if you intend to devote your life to scientific pursuits, why don't you first supply yourself with white paper and black ink.

Steinhaus himself valued his collaboration with Banach. He, Banach's discoverer, first teacher, and at times a guide, credited himself with only a modest role as an outside observer. We know, however, that a number of critical initiatives came from him: the founding of the Mathematical Society in Cracow, and the journal *Studia Mathematica*, joint lectures, and joint research discoveries. But there was something wonderful about Steinhaus' psychological makeup: he was not envious, he would help people without any ulterior motives, and he was himself modest. He was an ideal candidate for collaboration with almost anybody. In his autobiography he writes:

> In 1927, a collaboration with Banach resulted in the paper "Sur le principe de la condensation des singularités" published in *Fundamenta Mathematicae 9*. Henkel expressed the idea some time ago as a heuristic observation. Stanisław Saks helped edit the paper and later deepened the result by introducing in its proof the notion of category. This helped make the paper an important contribution to the Polish success between the two wars in the area of functional operations....
>
> In 1928 we [Banach and Steinhaus] founded the often-quoted *Studia Mathematica*. The first volume appeared in Lvov in 1929 – by now 19 volumes have been published. The journal is devoted to functional operations and should be considered as an organ of the so-called Lvov school. Until the war *Studia* was published in French, German, English, and Italian; after the war, Italian was replaced by Russian.

$$* \quad * \quad *$$

In the first volume of *Studia Mathematica* Banach published his paper "Sur les fonctionelles linéaires" in two parts. The paper contained extremely important results. The first part gave a proof of the fundamental functional-analytic result on extensions of linear functionals and their simple consequences. Today the theorem bears the name the Hahn-Banach theorem. The German mathematician Hans Hahn had proved it earlier, but Banach, who did not keep track of what was going on in the literature, did not know this. The second part contained a whole series of theorems and notions important for functional analysis. Among other things, Banach proved a theorem about the weak, sequential compactness of a ball in a dual space, and a theorem asserting that a set on which each linear continuous functional is bounded has to be bounded itself.

Banach also introduced the general notion of an adjoint operator and gave its basic properties. From these facts he derived a beautiful theorem on linear equations that was very close to the closed graph theorem, the "third pillar" of functional analysis that Banach would prove a little later.

Banach's creative life flourished. He wrote other papers as well during this period, but the fact worth noting is that more and more he worked jointly on his problems with other mathematicians. How those papers came about, and the atmosphere and circumstances under which they were born, will be explained in the next chapter on the legendary Scottish Café.

Banach, with Kuratowski, wrote "Sur une géneralisation du problème de la mesure." The paper contains an important result in set theory: under the assumption of the *continuum hypothesis*, it is impossible to define a measure of every set on the real line so that the measure is countably additive and each one-point set has zero measure. The methods Banach and Kuratowski used to prove this important result are still often employed. Furthermore, several other important works grew out of that paper.

Together with Saks, Banach wrote "Sur la convergence forte dans le champ L_p." The paper addresses the problem of summability in abstract spaces. This gave birth to a class of spaces that are still actively studied and that now are called spaces with the *Banach-Saks property.*

Thanks in part to Steinhaus' influence, Banach's research and scholarly activities expanded into other areas. While working with Kuratowski, he became socially involved with people from the Lvov-Warsaw philosophical school, a group that included several distinguished logicians and philosophers. The school had been founded in Lvov before World War I by people associated with the influential Cracow philosopher Twardowski. Later on, the school attracted such scientists as Jan Łukasiewicz, among others. The group was characterized by programmatic anti-irrationalism, a disdain for philosophical speculations, precision in expression, use of clear and rigorous language, a tendency towards broad utilization of mathematical logic, and adherence to the method of semantic analysis in philosophical studies. The school also had great influence on the development of the Polish school of mathematical logic. Banach was a close friend of one of its main representatives, Leon Chwistek.

Banach always spoke with admiration about Chwistek's papers, and when at some point Chwistek applied for a position in logic in Lvov, Banach backed him unequivocally and helped him to obtain the post. The affair scandalized half of intellectual Poland since Chwistek, in addition to being a respected scholar, also had a well-deserved reputation as being a somewhat strange and very eccentric artist.

*   *   *

Banach's work was by this time well known abroad. Banach himself participated in numerous international mathematical meetings. From September 3 to 10, 1928 he attended the International Mathematical Congress in Bologna. The Congress attracted over 800 mathematicians from 40 countries, including a sizable group of Poles, among whom were Sierpiński, who was in the presidium, Banach, Zygmund, Steinhaus, Nikodym, Leja, Chwistek, and Kuratowski.

At the Congress, representatives of the Slavic countries decided to convene in Warsaw a Congress of Mathematicians of Slavic Countries as an expression of the pan-Slavic sentiments that kept resurfacing periodically in Central and Eastern Europe. The implementation of this decision was entrusted to an executive committee comprising distinguished scholars from different countries, and the Congress was held in the fall of 1929. Inaugurated with pomp and circumstance on September 23, it attracted about 200 people. Five sections deliberated in the Aula [main auditorium and convocation hall] of the Warsaw Polytechnic:

I. The foundation of mathematics, history, and didactics,

II. Arithmetic, algebra, and analysis,

III. Set theory, topology, and their applications,

IV. Geometry,

V. Mechanics and applied mathematics.

Polish authorities supported the Congress and organized a number of official receptions. Prime Minister Świtalski and Minister Czerwiński

received the participants in the ornate chambers of the Presidium of the
Council of Ministers, and the city government hosted a banquet in the
City Council reception rooms. According to the Warsaw press, scientists
also attended an opera at the National Opera House and a banquet at the
Oaza restaurant.

The *Chronicle of the Jan Kazimierz University* lists invited guests
and other attendees:

> The Congress of Mathematicians of Slavic Countries was held on
> September 23–27, 1929, in Warsaw, under the honorary protectorate
> of the President of Poland, and was chaired by Professor Dr. Wacław
> Sierpiński. The Congress was attended by representatives of Bulgaria,
> Czechoslovakia and Yugoslavia; a number of Russian emigrés also at-
> tended. There were no visitors from the USSR since the Soviet govern-
> ment decided not to allow its citizens to participate in regional meetings
> that were based on nationalistic criteria. From the non-Slavic coun-
> tries, guests from Rumania, the German Reich, Austria, Holland, and
> Japan participated. W. Young from England, President of the Interna-
> tional Mathematical Union, was a guest at the special invitation of the
> Congress.

Among the next series of Banach's papers, some were written jointly
with his friend Stanisław Mazur. It should be noted that, just as Ruziewicz helped polish Banach's French-language papers, Mazur provided
the same services for papers written in German. Thus there appeared
"Über einige Eigenschaften der Lakunären trginometrischen Reihen"
and "Bemerkung zur Arbeit: 'Über einige Eigenschaften der Lakunären
trigonometrischen Reihen,'" where very general methods of functional
analysis were applied to problems on the theory of lacunary trigonometric series, a popular subject at the time.

* * *

In the spring of 1930 Banach won the Scientific Award of the City
of Lvov. This honor, named after the explorer Benedykt Dybowski,
was awarded for achievements in the areas of mathematical and natural
sciences. Banach did not rest on his laurels.

In 1931, under the editorial supervision of Banach and Steinhaus
from Lvov, and Knaster, Kuratowski, Mazurkiewicz, and Sierpiński from

Warsaw, a new series of *Mathematical Monographs*, published with the financial support of the National Culture Foundation, began to appear. Kuratowski believed that the creation of the new series was a turning point for mathematics in Poland. Here is his description:

> One of the most important events for Polish mathematics was the creation in 1931 of the *Mathematical Monographs* series. This event marked a new stage in the development of the Polish mathematical school. Earlier work which we could call the "pioneer" stage was characterized by the exclusive production of short papers containing new results (and published mainly in *Fundamenta Mathematicae* and *Studia Mathematica*). The time has arrived, however, for a great synthesis of the achievements of Polish mathematicians, or even a synthesis of all mathematical disciplines to which Polish mathematicians have made particularly significant contributions. The initial plan envisaged publication of monographs in the areas of functional analysis (Volume I – *Opération linéaires* by Banach), the theory of the integral (Volume II – *Théorie de l'integrale* by Saks), topology (Volume III – by this author), the continuum hypothesis (Volume IV – by Sierpiński), and the theory of orthogonal series (Volume V – by Steinhaus and Kaczmarz).

This early plan and more – *Trigonometrical Series* by Zygmund was added as Volume V, and Kaczmarz and Steinhaus' monograph appeared as Volume VI – was realized in an amazingly short period. All six volumes were published by 1935, and all have become true classics. Banach's *Theory of Linear Operations* met with a widely enthusiastic response soon after its publication. Americans were among the first to react. J. D. Tamarkin, discussing Banach's work and influence in the *Bulletin of the American Mathematical Society*, placed him in the pantheon of the giants of analysis: Volterra, Fredholm, Hilbert, Hadamard, Frechet, and Riesz. He wrote, *"Theory of Linear Operations* is fascinating in its own right, but its importance is further emphasized by numerous and beautiful applications."

Actually, Banach's monograph had appeared initially in 1931 in a Polish edition published by the Kasa im. Mianowskiego. Its title, *Teoria operacyj. Tom I. Operacje liniowe*, indicated that Banach's initial intention was to go beyond the standard linear functional analysis to nonlinear operators. The proposed second volume never appeared, however, although later on Banach and his collaborators Mazur and Orlicz developed a theory of polynomial operations. Banach dedicated the work to his wife. The French edition, which appeared in 1932 as the first volume of the *Mathematical Monographs* series, was dedicated more formally "A Madame Lucie Banach." This slim paperback of 256 pages cost 3 U.S. dollars, or 17.50 złotys for members of the Polish Mathematical Society.

According to historians of mathematics, this treatise contributed decisively to popularizing Banach's achievements among the general international mathematical community, and to the further development of his functional analysis.

Banach included in *Theory of Linear Operations* many results from the existing papers on functional analysis, and added many, many new theorems and applications. The theory that emerged from this work is very elegant and had a remarkable potential for applications. It was a work of international importance, and for many years mathematicians all over the world learned functional analysis from it. To this day the treatise contains quite a lot of interesting material that has not been outmoded by more modern works. From its various chapters whole branches of modern functional analysis developed. We can state without exaggeration that this work opened gates leading to broad and very attractive new mathematical territory.

The year 1931 saw the publication of another four papers by Banach.[2] The third used the category method to generate an extremely simple proof of the existence of a continuous function that does not have a derivative at any point. This fact was known to Weierstrass much earlier, but he had obtained his result through complicated constructions. The fourth paper on the Hölder condition, written jointly with Auerbach, generalizes the latter result to some extent.

* * *

The Second Congress of Polish Mathematicians was held September 23–26, 1931, in the Aula of the Stefan Batory University in Wilno. Forty-seven talks and six general addresses were delivered. Banach was invited to give one of the plenary addresses, and spoke on "Problems of the theory of vector spaces."

Banach was exceedingly busy during the early 1930s. In 1932 he began a three-year term as Vice-President of the Polish Mathematical Society. Also, his teaching duties at the Jan Kazimierz University and the Polytechnic continued unabated.

Banach's name appears several times on the roster of professors at the Lvov Polytechnic. He taught in the spring and fall semesters in the Faculty of Civil Engineering, as well as the General Faculty for academic years 1929–30, 1930–31, 1931–32, 1932–33, and 1933–34. The courses included theoretical mechanics (he taught this several times), calculus of variations, functions of several variables, and the theory of operations.

In 1934–35 Banach's name does not appear in the schedule of courses at the Lvov Polytechnic. Nikliborc took over his course in mechanics for geodesy students in the Faculty of Civil Engineering and the General Faculty was eliminated. The elimination of the General Faculty was a result of the so-called Jędrzejewicz's reforms. For reasons never clearly explained, the Minister decided to abolish the General Faculty, which had been created by the former Prime Minister.

Between 1932 and 1936 seven new papers by Banach appeared, several of them written jointly with coauthors such as Mazur and Kuratowski. These papers continued to build on the foundations of functional analysis established in the *Theory of Linear Operations* and broadened the trend of applying the discoveries of functional analysis to other areas.

*   *   *

By 1936 Banach was already an author of some 47 papers, and was invited to deliver a one-hour plenary address at the International Mathematical Congress in Oslo, a major honor. He chose to speak on the "Theory of operations and its significance in analysis." Antoni Zygmund, then at the Stefan Batory University in Wilno and later Professor at the University of Chicago and father of the American school of harmonic analysis, wrote about the Congress in the 1938 issue of *Mathesis Polska*:

The Oslo Congress was not as well attended as the previous ones. The number of participants was only about 500 compared to 1200 in Bologna and 700 in Zurich. The drop-off was partly due to currency restrictions imposed by several countries that made it more difficult to travel to meetings. The political climate was also a factor. The Italians categorically refused to participate, blaming their refusal on international sanctions imposed on Italy. The Russians did not show up either, although they originally promised to attend. At the last minute, they were stopped from coming by issues related to an impending major reorganization of Soviet mathematics. Even the Germans, normally numerous and eager to participate, failed to show up. From the Reich, only about 20 came. The small number of participants from Germany may have been in retaliation for the fact that the University of Oslo did not send a delegation to the anniversary celebrations at Heidelberg University held earlier this year.

... Americans formed the largest group at the Congress with about 70 mathematicians. The Polish contingent included Borsuk, Eilenberg, Grużewska, Lubelski, Sierpiński, Straszewicz, Zarankiewicz (all from Warsaw); Banach, Kaczmarz, Schauder, Żylinski (all from Lvov); Gołąb, Ważewski, Zaremba (all from Cracow) and Zygmund (from Wilno).

Despite the smaller number of participants, the Congress was a success because of excellent organization and the topics discussed. Zygmund recalled:

> It became a custom for two kinds of talks, plenary and sectional. The number of plenary talks is always limited; they occur in the morning and usually last for about an hour. Organizers assign these addresses only to the most outstanding mathematicians who talk about more general topics but with a heavy inclusion of their own work. An invitation to deliver such an address is always considered a great honor. Among Polish participants such an address was delivered by Banach.

In 1937 Banach's "The Lebesgue integral in abstract spaces" was published as an appendix to the new English version of Saks' *Theory of the Integral*. It contained a nonstandard approach to the theory of the integral and many original ideas were not unlike those that had appeared somewhat earlier in the work of the American mathematician P. J. Daniell. In 1938 Banach wrote "Über homogene Polynome in

(L^2)," which opened up a new area of so-called polynomial operations in functional analysis. Mazur and Orlicz developed this direction further. There had been a renaissance in the field in connection with the probabilistic theory of multiple stochastic integrals.

In the last academic year before World War II Banach published only one paper, "Über das 'Loi suprème' von Hoene-Wroński," but also started work on "Sur la divergence des séries orthogonales" and on "Sur la divergence des interpolations," which contained a number of interesting observations on approximation theory. He also wrote a textbook, *Mechanics for Academic Schools*, which was quite popular and enjoyed a good reputation. In 1951 it was translated into English as the 34th volume of the *Mathematical Monographs* series.

$$* \quad * \quad *$$

Just before the war, Banach received a number of honors. In April 1939 he was elected President of the Polish Mathematical Society, the crowning achievement of his active involvement in the society he helped to found twenty years earlier. The Society, by now well established, also acted as a conduit through which many distinguished foreign mathematicians arranged their working visits with Polish mathematicians. Among those who visited Poland were Pavel Aleksandrov, Emil Borel, Elie Cartan, René Maurice Fréchet, Harold Hardy, Leonid Kantorovich, Henri Lebesgue, Solomon Lefschetz, Tulio Levi-Cività, Nikolai Luzin, Karl Menger, Paul Montel, Frigyes Riesz, Ivan Vinogradov, and Ernst Zermelo.

On June 9, 1939, the General Assembly of the Polish Academy of Knowledge, using resources made available by the Janina Ryczter Mościcka Foundation, awarded Banach the Grand Prix in the amount of 20,000 złotys (about \$5,000) – a considerable sum, close to a typical annual salary of an American university professor in those days – for his paper "Sur les fonctionelles linéaires." It was the first such prize given by the Academy in the areas of mathematics and astronomy. It was also the highest monetary award given in Poland in any field. The sum was deposited in Banach's account but he was never able to access it – with the outbreak of the war, all bank accounts were frozen. The official award ceremony was to take place in October, the beginning of the new academic year, but the war started on September 1 and no celebration was possible. By October 1 Soviet troops had occupied Lvov.

CHAPTER V

The Scottish Café

WHAT THE CAFÉS OF MONTMARTRE did for the arts of Fin-de-Siècle Paris, the Scottish Café did for mathematics in Lvov. It was a "crazy" but "sacred" place. In 1936 the Polish mathematician Edward Marczewski (1907–1977)[1] claimed:

> Poland has always had great individuals who often successfully did the work of many, sometimes for whole institutions, sometimes for whole generations. Today, however, Poland's mathematical community is not only a collection of distinguished individuals, but also a numerous and well-organized team totally devoted to creative research.

Given the strong nature of Polish individualism, it has been an ambitious undertaking to explain the incredibly fruitful collaboration of a group of unusually gifted and original minds. By examining the life of Lvov's Scottish Café, we get some insight and a partial answer as to how this collaboration worked.

The atmosphere of Lvov must have played a role in developing the mathematical circle at the Scottish Café. In the book *Old and New Lvov*, published in 1928 by Małopolska Agencja Reklamowa, Michał Lityński gives an admiring, if somewhat exalted, description of his city [akin to what was described in the earlier chapter on Lvov]:

> At night, when blinding illuminations light up the windows of the fine department stores, and impenetrable crowds fill the wide, gas lantern-lined sidewalks, alongside the lawns and flower beds lovingly cared for by the city, when the street pavement resonates with the deafening noise of speeding cars, buses, streetcars and the slowly disappearing horse carriages, then one can easily forget that one is only strolling through the capital of one of the Polish provinces. Also, we must mention the universal grace, beauty and chic fashions of the Lvov women. It is pleasant to live in such a city and, even for a moment, become immersed in its exuberant Polish life. Not surprisingly, the fame of our city continues to attract a large number of out-of-town visitors who afterwards leave with most pleasant memories.

Judging by the testimonials of other mathematicians, Lityński's evaluation corresponded to reality. The elemental joy and accumulated energy released by Poland's reclaimed independence – to use Zygmund's expression – would draw people out of even the most spacious apartments into the public squares, wide streets and promenades created in place of the old city walls and ramparts.[2]

The neighborhood of Academicka Street and Akademicki Square – a couple of hundred yards from the university building – was at that time one of the most frequented districts in Lvov. At 22 Akademicka Street was Ludwik Zalewski's Confectionery, patronized by Łomnicki, Steinhaus and Kuratowski. Zalewski's Confectionery was one of the most elegant establishments in Lvov; at noon and in the evening it overflowed with people. Legend had it that the café served the best pastry in Poland. Almost as good were the Welz Confectionery at 5 Akademicka Street, Dudek's at Mariacki Square, the Europejska at 5 Hetmańska Street, and the confectioneries of Urbanik and Lewandowski. The Café Roma was also on Akademicka Street, and was frequented mostly by the army officer corps and the intelligentsia. And across the street was the legendary Scottish Café.

*   *   *

Initially, mathematicians would meet at the Café Roma on Saturday night, following the weekly meetings of the local chapter of the Polish Mathematical Society, which were held in the University's Mathematics Department seminar room. The meetings usually included four or five ten-minute talks; longer talks were rare. The discussions then overflowed to Café Roma. Banach started joining them, almost on a daily basis, but soon became irritated with the establishment's handling of his credit situation and decided to move the gatherings across the street to the Scottish Café.

The 1925 edition of *The Illustrated Guide to Lvov* notes that the Café was located at 9 Akademicki Square, that it was owned by one Mr. Zieliński, that it was a meeting spot for sports, literary, and university figures, and in the evening for good music. The author of the guide did not specifically mention the mathematicians since in 1925 their fame had not yet been established.

The Scottish Café was decorated Viennese style: tiny tables with marble tops were extremely useful as tablets to be covered with mathematical formulas. At first, the owner was not overly enthusiastic about such activity. However, after a while Zieliński got used to this "destruction" of his property, at least until Mrs. Banach mercifully rescued him by purchasing a ruled notebook, which was to become the celebrated *Scottish Book*. After all, the tables were occupied not by teenage troublemakers, but by serious professors of the university and polytechnic.

The meetings of mathematicians at the Scottish Café were initially irregular – but pretty soon a certain rhythm and custom were established, as daily sessions at the Scottish grew into an enduring ritual among many mathematicians. Frequent initial participants included Banach, Stożek, Ruziewicz, Steinhaus, Kaczmarz, Żylinski, and the colleague to whom Banach was closest, Mazur. Later the circle expanded to include Nikliborc, Auerbach, Schreier, Schauder, Kuratowski, Nikodym, Ulam, Eilenberg, Orlicz, Eidelheit, Kac, and Birnbaum. Also, later, Józef Marcinkiewicz arrived from Wilno as Banach's postdoc. Usually they began arriving between 5 and 7 P.M. – always occupying the same tables – and for the next several hours they worked with total concentration, covering the marble tabletops with mathematical formulas. But saying that "they worked with total concentration" is not completely accurate, as there was no meeting without jokes, heated discourse, shouting, and drinking. Banach, for example, drank enormous amounts of coffee and cognac and smoked dozens of cigarettes.

Nor were the discussions purely mathematical. Stożek, for example, often played chess with Nikliborc, and other mathematicians drank coffee or cognac and cheered them on. Auerbach was also an avid chess player. No one was bothered by the mediocre food and cognac, and both were consumed in huge quantities. Ulam's memoirs recount the mathematical and nonmathematical side of the Scottish Café evenings:

> Kuratowski and Steinhaus appeared occasionally. They usually frequented a more genteel tea shop that boasted the best pastry in Poland.
>
> It was difficult to outlast or outdrink Banach during these sessions. We discussed problems proposed right there, often with no solution evident even after several hours of thinking. The next day Banach was likely to appear with several small sheets of paper containing outlines of proofs he had completed. If they were not polished or even not quite correct, Mazur would frequently put them into a more satisfactory form.
>
> Needless to say such mathematical discussions were interspersed with a great deal of talk about science in general (especially physics and astronomy), university gossip, politics, the state of affairs in Poland; or, to use one of John von Neumann's favorite expressions, 'the rest of the universe.' The shadows of coming events, of Hitler's rise in Germany, and the premonition of a world war loomed ominously.

The cast of characters at the Café was very colorful. Stożek, whose cheery disposition and permanent good mood was infectious, was the Dean of the General Faculty at the Polytechnic. Although his name means "a cone" in Polish, he looked more like a ball – rotund, small, and completely bald. He was addicted to hot dogs smothered in horseradish, which he claimed cured melancholy. Kaczmarz was tall and slender and Nikliborc – short and squarely built; they were often seen together. They reminded Ulam of the then popular movie comedians Pat and Patachon.

Bronisław Knaster was a kind of amateur mathematician with a Sorbonne degree in medicine who, incidentally, specialized in the construction of pathological sets in topology. His sense of humor was proverbial and he liked to entertain the group with examples of hilarious *volapük* from multilingual and multicultural Lvov students. Once he overheard a conversation between two students in a restaurant: "*Kolego, pozaluite mnia ein stuckele von diesem faschierten poisson,*" an amalgam of Polish, Russian, Ukrainian, Yiddish, German, and French, all in one short sentence.

A humorous anecdote about Knaster was preserved and quoted by Gołembowicz in his collection *The Scholars in Anecdotes*:

> Stanisław Saks, a Warsaw mathematician on a stroll with Knaster, noticed a shop window with the caption: "hare paté."
> – What do you think – he skeptically asked Knaster – is this a genuine hare paté?
> Knaster replied:
> – Whoever eats it will know right away. *Jesli nie zajęczy to zajęczy, a jesli zajęczy to nie zajęczy.*

This is a nearly untranslatable play on two Polish words: *zajęczy* means both "made of hare" and "he will moan," although the etymological roots are quite different. Thus, the literal translation is: "If he moans not then it is made of hare, and if he moans then it is not made of hare."

This sense of humor was, in fact, one of the most important features of the Lvov mathematical community. Anecdotes circulated by the dozen. Here is another one: One spring, Herman Auerbach, Docent of Mathematics at Lvov University, bought himself a new fedora. Shortly thereafter somebody took it from the Café, leaving in its place a similar but well-worn specimen. Unfazed, Auerbach took to wearing this foundling hat without ever bothering to clean it. Asked why he chose this course of action he responded, "I'm not going to clean the thief's hat."

Other funny stories about Auerbach also provide an extra insight into daily life at Lvov University. When someone complained about the disorder reigning in the mathematics library, he noted that "chaos is better than order. Although amid chaos you may not be able to find anything, you cannot lose anything either."

Everybody felt at ease in Lvov, even the foreigners. An episode during Henri Lebesgue's 1938 visit at Jan Kazimierz University to receive the honorary doctorate – arranged by Steinhaus, who among Polish mathematicians knew Lebesgue best, and was at that time Dean of the Faculty – was described in Marek Kac's reminiscences:

> At the time of his visit Lebesgue was no longer interested in anything but elementary mathematics; he refused to discuss measure, integrals, projection of Borel sets or anything of that sort. He gave two lectures, both extremely beautiful, but entirely elementary: one on construction by ruler and compass, and the other on iterated radicals....
>
> ... We had a 5 o'clock reception for Lebesgue in the Scottish Café. Fewer than 15 people attended, which goes to show how small the number of mathematicians was in those days. The waiter gave all of us menus, and not realizing that Lebesgue was not a Pole, he gave him one too. Lebesgue looked at the menu for about 30 seconds with utmost seriousness and said, *"Merci, je ne mange que des choses bien définies."* [Thank you, I eat only well-defined things.]

Another colorful personality, at the other end of the spectrum of characters moving in the Scottish Café circles, was a high school science teacher by the name of Hirniak. Small in stature but with a great imagination, he tried to work on the Fermat conjecture, which until recently had been one of the most famous unsolved problems in mathematics. Usually, he stayed at side tables at the Café, intermittently sipping vodka and coffee and scribbling things on a piece of paper. Nikliborc nicknamed him "Gehirniak" (*Gehirn* means brain in German). Hirniak was completely unaware of the humor he generated, and his statements circulated around Lvov, including the time he explained to Banach his proof of the Fermat conjecture and afterwards added: "The longer and larger my proof, the smaller the hole in it."

He wildly exclaimed that American reporters would one day discover his proof and would descend on Lvov, asking:

– Where is this genius? Give him one hundred thousand dollars!
And Banach, amused, would echo:

– Yes, give it to him!

Ulam, who repeated this story in his memoirs, said that it amused von Neumann to no end when the Defense Department started distributing grants of similar size after the war. He used to say:

– Remember how we laughed at Hirniak's hundred thousand dollars story? Well, he was right, he was the real prophet while we laughed like fools. Not only was he right, but he even foresaw the correct amount!

Usually students did not participate in the Scottish Café meetings, and only two friends, Stanisław Ulam and Józef Schreier, were admitted to the august gatherings during their student years. Professor Andrzej Alexiewicz, a functional analyst at the University of Poznań (recently deceased), claimed that an invitation to the sessions at the Scottish Café was akin to being knighted.

Visitors from Wilno, Warsaw, and Cracow who tasted the atmosphere of the Lvov mathematical community had a hard time leaving. Kazimierz Kuratowski, a lifetime Warsaw resident, wrote about the few years he spent in Lvov after being appointed to a chair there:

> I consider my stay in Lvov as a particularly happy event. The warm collegial atmosphere of the Lvov mathematical community was enhanced by the special charm of an unusual city. For me personally, the closer acquaintance and active collaboration with Stefan Banach, the most famous Polish mathematician, was extremely valuable. Working in Lvov while maintaining frequent contacts with the Warsaw community was the most fruitful period of my scientific career.

*   *   *

Of course, mathematics was the main topic at these gatherings. Ulam remembers them well:

> There would be brief spurts of conversation, a few lines would be written on the table, occasional laughter would come from some of the participants, followed by long periods of silence during which we just drank coffee and stared vacantly at each other. The Café clients at neighboring tables must have been puzzled by these strange doings. It is this kind of persistence and habit of concentration which somehow emerges the most important prerequisite for doing genuinely creative mathematical work.

Steinhaus acknowledged that the table at which Banach sat with Mazur, and later also with Ulam, was center stage at the Scottish Café.

Their sessions were convened daily. There was one that lasted 17 hours. Its result was a proof of a certain important theorem about Banach spaces, but nobody put it on paper. Today nobody knows how to reproduce it, because – as Steinhaus wrote – after the closing of the Café, the marble tabletop covered with the formulas in "chemical pencil" was, as usual, wiped clean by the janitor. That was the fate of many a theorem proved by Banach and his disciples. Banach's work style awed Steinhaus:

> ...as a witness of Banach's work, may I be permitted to say that he had a clarity of thinking that Kazimierz Bartel once called "almost unpleasant." He never counted on a lucky turn of events, or that the desired assumptions be satisfied just when needed, and often would say that "hope is the mother of fools"; he applied this disdain for optimism not only to mathematics but also to political prophesies. He was similar to Hilbert in the way he attacked problems directly – after excluding by examples all the sidetracks, he concentrated all his efforts on the remaining road leading directly to the goal – he firmly believed that the logical analysis of a problem, conducted in the way a chess player analyzes a difficult position, has to lead either to a proof of the theorem or to a counterexample....

But he also wrote that:

> We have to regretfully state here that many valuable results of Banach and of his school were lost, with a serious loss to Polish science, as a result of lack of pedantry among members of the school and, first of all, Banach himself.

Andrzej Turowicz offered this rejoinder:

> Unfortunately, that's not an exaggeration. I believe more was lost than published. Of course, Banach did publish the most important things, but not all of them. I can explain the reasons. Banach was not able to write down and edit the theorems that he was producing at a stupefying rate. People imagine that creative mathematical work consists of sitting behind a desk and doing calculations. Nothing is further from the truth. The job usually is to answer the question whether a certain theorem is true or false, and one has to come up either with proof of its validity or a counterexample to the statement under consideration. Sometimes this sort of contemplation takes many hours over many weeks or months. I know the case of a mathematician who stubbornly was thinking about a certain problem for a dozen or so years. Here,

Banach broke all the records with the speed by which he obtained new results. And solving one problem most often leads to another problem. So Banach never made notes of his scientific discoveries. He just communicated them orally to the Lvov community. If two or three of them ever got down to editing what they had heard from Banach, his whole oeuvre would have been saved. But as it was, everybody preferred to concentrate on his own creative work rather than write down somebody else's theorems. If the tape recorder had been known during Banach's life, then we could have recorded his whole work on tape. But he did not live long enough to see tape recorders in use.

It was said and written in Poland and abroad that the café work culture constituted the "Polish way of doing mathematics," a phenomenon of teamwork in unorthodox places that led to joint solutions of research problems. Banach is usually given credit for creating this working style. Ulam further comments:

> ... In mathematical discussions, or in short remarks he made on general subjects, one could feel almost at once the great power of his mind. He worked in periods of great intensity separated by stretches of apparent inactivity. During the latter his mind kept working on selecting the statements, the sort of alchemist's probe stones that would best serve as focal theorems in the next field of study.
> ... The breadth of his cooperation and his intensity were rare traits which I have never encountered elsewhere except during the war years at Los Alamos.

This intensity is obvious when one peruses pages of the *Scottish Book*.

* * *

The Scottish Book: What was it and how did it come about?

Today, it is one of the most revered relics of the mathematical world. As with any legend, some of the details of its history differ depending on the person recalling them. It started out as simply a regular, ruled school notebook with a marble-patterned cardboard cover, for which Banach's wife Łucja paid two and a half złotys at a drugstore. Evidently she was disgusted with her husband and his friends' favorite activity: smearing computations and mathematical problems on the Café's tabletops. The notebook was kept in the Café's cloakroom and given to mathematicians on demand, although Steinhaus claimed that the guardian of the *Scottish Book* was neither the cloakroom attendant, nor the waiter, nor the owner, but the cashier of the Café. He also stated:

> The problems in the *Book* would be written on the odd-numbered pages of consecutive sheets, with the opposite pages left blank for possible future answers.

Anyone interested could write down problems to be solved and anyone could write his solutions. In a sense, it was an unofficial communal scientific publication. Kuratowski wrote the following about the founding of the *Scottish Book*:

> During numerous meetings in the Scottish Café (the favored café of Lvov mathematicians), the number of posed new problems grew so fast that at some point it was decided that it made sense to write all these problems in a special notebook, to be kept permanently at the Café. Thus, the legendary *Scottish Book* was created. It had a substantial scientific, emotional, and historical value, because of the names of the contributors, often distinguished foreigners.

In the first problem entered in the *Book* on July 17, 1935, Banach asked about conditions of metrizability of (what we now call) a Banach space as a compact and complete space. By 1941, when the *Book* was closed, 193 problems had been recorded by several dozens Polish and foreign mathematicians. They were not evenly distributed among the participants in the Scottish Café ritual. Banach offered 14 by himself (plus another 11 jointly with Mazur and Ulam), Ulam 40 (plus 15 jointly), Mazur 24 (plus 19 jointly). These three clearly dominated the proceedings. Steinhaus entered 10, and the rest of the regulars, Ruziewicz, Auerbach, Kac, Eilenberg, Orlicz, Nikliborc, and Schreier averaged about half a dozen each. And then there were scattered entries from distinguished invited visitors: Frechet, Zygmund, Offord, Kampe de Feriet, von Neumann, Sobolev, and Lusternik.

The contributions were also not evenly distributed over time. Out of the 193 problems in the *Book*, 122 were entered in the first six month of its existence, and then things tapered off. The year 1936 brought a collection of 32 problems,[4] 1937–13, 1938–9, 1939–4, 1940–7, 1941–4. Clearly, as Ulam writes,

> Numerous problems came about from before 1935; they had to be thoroughly analyzed before inclusion by those over whose names they appeared. The majority of proposed questions had to be discussed in depth before they would be considered for official entry. In some cases the problems would be resolved on the spot and answers written in right away.

In the few years of its active use, the fate of the *Scottish Book* reflected the history of the City of Lvov itself. Shortly after the beginning of World War II on September 1, 1939, the city was annexed by the Soviet Union, and entries of Soviet mathematicians such as Bogolubov, Aleksandrov, Sobolev and Lusternik appearing in the *Book* indicate the acute interest of the new powers in the workings of the Lvov mathematical school. The last problem in the *Book* was entered by Steinhaus on May 31, 1941, and contained a rather puzzling collection of numerical results concerning partitions of matches in a matchbox! After the beginning of the hostilities between the Reich and the Soviet Union, and the occupation of the city by German troops in the summer of 1941, the entries ceased.

Anticipating the war, some mathematicians were thinking about safe-guarding the *Book*. According to Ulam:

> ... in the Summer of 1939, during my last visit in Lvov and a few days before my return to the United States, I talked to Mazur about the probability of war.[5] People in general expected another crisis in the style of Munich and were not prepared for the coming World War. Mazur said:
> "The possibility of a World War is real. What are we going to do with the *Scottish Book* and our joint, unpublished papers? You are going to the United States, and most surely you are going to be safe. In case the city is bombed, I will pack the manuscripts and the *Book* in a chest and bury it."
> We even decided on an exact place – next to a goal post at the soccer field on the city outskirts. I do not know if that's what happened. In any case, the manuscript of the *Scottish Book* survived the war in good shape. After the war Steinhaus sent me a copy. In 1957 I translated it into English and sent it around to many friends – mathematicians in the United States and elsewhere.

At the 1958 International Mathematical Congress in Edinburgh, Ulam made a photocopy of the *Scottish Book* available to its many participants. Because of its name, it initially created a great sensation among the hosting Scots, who were disappointed to learn that the Scottish connection was purely nominal.

Many problems from the *Scottish Book* played a significant role in the development of functional analysis and other branches of mathematics. Mazur's problem (number 153, dated November 6, 1936) about the existence of Schauder bases in separable Banach spaces remained one of the central open problems of functional analysis until, in 1972, a Swedish mathematician Per Enflo (now at Kent State University in Ohio) solved it in the negative. Mazur offered a prize – a live goose – to the solver, and the prize was duly given to Enflo when he visited Warsaw to give lectures on his solution. It was a major media event. Many other prizes were offered in the *Book* for solutions to the problems announced. Again, Mazur was most active in this department but, with the exception of the goose, his prizes were mostly pedestrian, a few bottles of beer or a bottle of wine. The ever-patrician Steinhaus offered prizes only twice, but it was 100 grams of caviar on the first occasion and a dinner at the posh George's on the second. Banach offered only one prize (a bottle of wine).

Foreign visitors chipped in with fancy (but perhaps hard to collect) prizes. Von Neumann offered a "bottle of whiskey of measure > 0," Ward a lunch at Dorothy's in Cambridge and Wavre a fondue in Geneva. It is not clear if travel expenses were covered. The Soviet conquerors offered celebratory spirits: Bogolubov a flask of brandy, Sobolev a bottle of wine, and Lusternik champagne; but by that time the local population, occupied and starving, was in no mood to celebrate and turned practical. On a gloomy and freezing February 8, 1940, Stanisław Saks offered one kilo of bacon for solving a problem on subharmonic functions.

After the war, the original of the *Scottish Book* ended up in possession of Banach's wife Łucja, who brought it to Wrocław. After Łucja Banach's death in 1954 (she was buried at a Wrocław cemetery under the gravestone caption "Mathematician's wife") the *Scottish Book*'s ownership passed into the hands of Banach's son, Stefan Jr. (born October 14,

1922), a Warsaw neurosurgeon, in whose possession it is today. In the 1980s he put it briefly on display at the Banach Center of the Institute of Mathematics of the Polish Academy of Sciences in Warsaw. In 1981, Birkhäuser published its printed version, edited (with considerable assistance from Jan Mycielski of the University of Colorado at Boulder) by Daniel Mauldin, a mathematician from Texas, and complemented with extensive mathematical and historical comments on individual problems.

Steinhaus, who settled in Wrocław, prodded somebody there to buy a new notebook. It was named the *New Scottish Book*. It has been in use since 1946, with its contents regularly published in *Colloquium Mathematicum*, a scholarly journal founded by Edward Marczewski. It fulfilled a role similar to that of the original *Scottish Book* in Lvov – minus Banach and the Scottish Café. Initially it was kept in the joint library of the Wrocław University and Polytechnic Mathematical Seminar, and, after 1970, when the University had a new building constructed for the Institute of Mathematics, was moved there. The tradition of the *Scottish Book* thus continued.

*   *   *

One may wonder how was it possible to conduct serious and creative scientific activity in the seemingly chaotic daily gatherings at the Scottish Café. But the Lvov situation was not unique; there are other historical figures whose contemplation, reflection, and intense mental effort were helped by the din, the noise, and the commotion of their surroundings. In Michel de Montaigne's *Essais*, we find an interesting comment on this subject:

> Not long ago I observed one of the wisest scholars in France – also a person of considerable wealth – as he devoted himself to his studies in the corner of a room, which was separated by a curtain from the area where the servants' frolics filled the air with din and commotion. He told me (and Seneca says almost the same thing) that such chaos does him good: deafened by the noise, he is forced to concentrate and focus inside himself for the purpose of contemplation, and the storm of voices somehow corrals and compresses his thoughts inside him. While a student in Padua, for a long time he occupied a room next to a campanile, and was exposed to the noise of passing carriages and

the nearby market. As a result he not only got used to the noise, but found it beneficial for his studies. Socrates' response to Alcybiades, who queried: "How can one bear the constant croaking of one's wife?" was "The same way as one can get used to the continuous creaking of a wheel drawing water." Things are very different for myself: my mind is very sensitive and easy to disturb and when I'm focused, the buzzing of the tiniest fly can destroy my concentration.

Clearly, different people with different sensitivities, personalities, and psychological makeups react differently to the same surroundings. So it was remarkable, and perhaps serendipitous, that in Lvov one assembled at the same time a collection of many mathematicians of compatible personalities who thrived on working in quite improbable circumstances. The collection of individuals gathered around Banach in the Scottish Café would probably be impossible to replicate in another time or under other conditions.

Antoni Zygmund, at the Stefan Batory University in Vilnius at that time (at the University of Chicago after the war) and an alumnus of the Warsaw mathematical school, offers an explanation of the origins of this phenomenon in an interview with Jerzy Jaruzelski, modestly disclaiming any competence here:

> ...So, you are asking me what was the soil in which this phenomenon grew? I will briefly say: intellectual freedom and accumulated energy released by reestablishment of the Polish independent statehood. And as always – Sir – hard work. Perhaps early, at the very beginning, spontaneous and impatient, but later on – quiet and systematic. Finally, of course, distinguished individuals.
>
> Remember that Polish mathematics existed also in the 19th century, although it was dispersed at universities of the occupying powers, especially Russia and Austria. More precisely, it was not so much Polish mathematics as Polish mathematicians. Let me recall the name of Antoni Przeworski at the University of Kharkow, Stanisław Zaremba who taught in France and later in Cracow, or Wacław Sierpiński. Hence the substance – and quite a solid substance it was – existed there earlier. However, the regaining of independence caused tremendous intensification of organizational and scientific activity. For sure, a lot of credit should be given to Zygmunt Janiszewski.[6]

Generally speaking, in the creation of the two schools, quite natural ingredients were involved: the knowledge that was brought by the elders, that is thirty-year-old colleagues, and young talents which the former managed to gather around themselves. Stefan Banach's genius was certainly buttressed – at least in its beginning stage – by the erudition and encyclopedic knowledge of Hugo Steinhaus. The situation was similar with others.

Village of Ostrowsko in the foothills of Tatra Mountains. Banach's father, Stefan Greczek was born here. *(Photography: WAW)*

From right to left: Banach's half-sister Antonina Waksmundzka, her daughter, grand-daughter and husband photographed in 1993 in front of a summer house Banach's father Stefan Greczek built in his native village of Ostrowsko in the 1930s. *(Photography: WAW)*

St. Nicolaus Church in Cracow where Banach was baptized in April of 1892.
(Photography: WAW)

Fourth grade students of the Cracow's Fourth Gymnasium in 1905/06.

A magnification of the fragment of the previous photograph. Marian Albiński [1], Stefan Banach [2], Witold Wilkosz is at the bottom.

Cracow's Planty promenade. In the spring of 1916 Steinhaus "discovered" Banach discussing Lebesgue measure with Otto Marcin Nikodym on one of these park benches. *(Photography: WAW)*

Jagiellonian University's Philosophical Seminary. Here, in 1919, Zaremba and others met to establish the Polish Mathematical Society. Young Banach was one of the signatories of the constitutional assembly. *(Photography: WAW)*

Stanisław Zaremba (1863–1942).

Young Wacław Sierpiński (on the left) and Włodzimierz Stożek (on the right).

Hugo Dyonizy Steinhaus, an earlier portrait.

Zygmunt Janiszewski in the uniform of the Polish Legion.

SUBDIVISION DES MATHÉMATIQUES

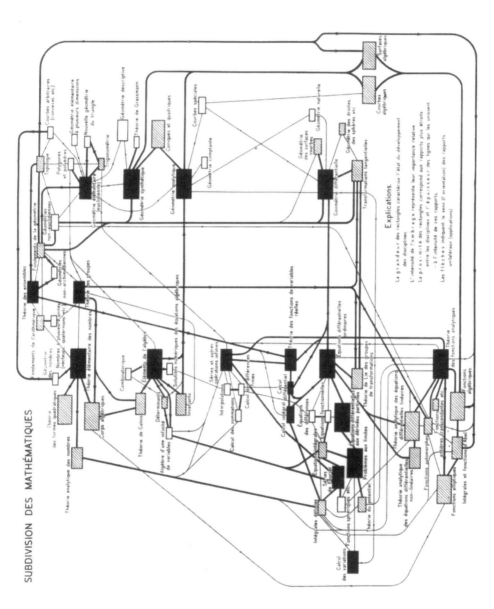

Janiszewski's 1918 "subdivision of mathematics."

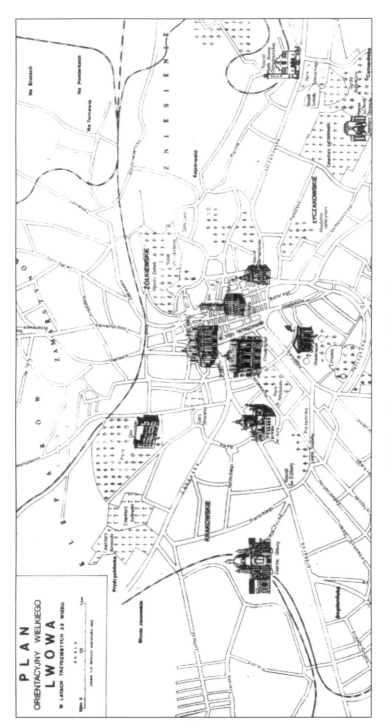

The City of Lvov street plan in the 1930s.

The old Jan Kazimierz University building adjacent to the St. Nicolaus Church. The mathematics seminar had offices on the second floor.

The Main Building of the Lvov Polytechnic. Banach studied there for two years beginning in 1910.

Akademicka Street in Lvov in 1930s.

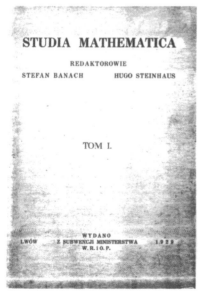

STUDIA MATHEMATICA

REDAKTOROWIE

STEFAN BANACH   HUGO STEINHAUS

TOM I.

WYDANO
LWÓW   Z SUBWENCJI MINISTERSTWA   1929
W.R.IO.P.

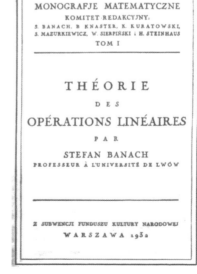

MONOGRAFJE MATEMATYCZNE
KOMITET REDAKCYJNY:
S. BANACH, B KNASTER, K. KURATOWSKI,
S. MAZURKIEWICZ, W. SIERPIŃSKI i H. STEINHAUS
TOM I

THÉORIE
D E S
OPÉRATIONS LINÉAIRES
P A R
STEFAN BANACH
PROFESSEUR À L'UNIVERSITÉ DE LWÓW

Z SUBWENCJI FUNDUSZU KULTURY NARODOWEJ
WARSZAWA 1932

Facsimile of the front page of the
first volume of *Studia Mathematica*

Facsimile of the title page of
Banach's *Théorie des opérations
linéaires*

Villa "Gerlach" in the Tatra Mountains resort town of Zakopane. It was owned by Leon Chwistek and his sister Anna Stożek, and around 1930 Banach, Sierpiński and Włodzimierz Stożek spent many a summer month there working on their school textbooks. *(Photography: WAW)*

Zermelo in Lvov (1930). First row (l to r): Steinhaus, Zermelo, Mazurkiewicz. Second row: Kuratowski, Knaster, Banach, Stożek, Żyliński, Ruziewicz.

The Scottish Café, birthplace of the *Scottish Book* problems, as it appeared in a postcard from the early 1970s. The Scottish Café is on the right, with the Café Roma on the left. Stanisław Ulam observed that the scene has changed little from the period preceding World War II. (©1981 Birkhäuser Boston)

Another view of the Scottish Café. (©1981 Birkhäuser Boston)

Facsimile of a page from the *Scottish Book*. (©1981 Birkhäuser Boston)

Włodzimierz Stożek wrote popular school textbooks, some of them jointly with Banach. His name means "cone" in Polish.

Mazur (on the left) and Ulam on a Lvov street around 1935.

Stanisław Saks as a young man.

Stefan Banach in the 1930s.

Kazimierz Kuratowski, a portrait
from 1970s.

Bronisław Knaster, a later portrait.

Stanisław Saks in a 1938 caricature by Jeśmanowicz.

Juliusz Schauder (1896–1943).

Władysław Orlicz, a later portrait.

Alfred Tarski, a portrait from his
later (Berkeley) days.

Stanisław Ulam after World War II.

Samuel Eilenberg, a later portrait.

Wacław Sierpiński (1882–1969), a later portrait.

Hugo Steinhaus, a later portrait.

Stefan Banach in 1944.

Stanisław Mazur hands the goose to Per Enflo at a 1972 Warsaw ceremony.
In 1936, Mazur promised the goose as a prize for solving a problem he posed
in the *Scottish Book.*

Stefan Banach's grave at the Łyczakowski Cemetery in Lvov. The Riedl family of
Lvov merchants accepted his remains in their tomb.

CHAPTER VI

Banach Privately and in Daily Life

W HAT WAS B ANACH LIKE in everyday life? From facts, relations, conjectures, and indirect evidence gathered to date by the author, a sympathetic image emerges of an authentic self-made man, without any signs of aggressiveness, a good colleague and friend; above all a charming companion, and a pleasant supervisor of young people.

To be sure, he had many weaknesses, but he led an authentic life, without much pretense or hypocrisy, at peace with himself. He paid no attention to artificially established canons. Banach's unconventional behavior shocked the staid academic community and the establishment. Anecdotes about what Banach did, how he entertained himself, how he dressed, what he said, whom he offended, and so on, circulate to this day in the mathematics community. For instance, it is known that the good citizens of the city of Lvov found it revolting that, during the summer, Banach would parade around in a short-sleeved shirt, pants supported by a belt instead of the then *de rigueur* suspenders, chomping on a cigarette holder, and carrying a massive cane. Banach simply did not fit the standard image of a proper professor and scholar. He also refused to don high hat and tails, then common in academic circles.

Pictures of Banach depict a well-built, tall, and broad-shouldered person, with a high forehead, smoothly combed parted hair, and a half-skeptical, half-ironic smile. What is most attractive in this face is the mesmerizing expression of his blue eyes: piercing and gleaming as they x-rayed the person with whom he was conversing, and yet he seemed to be gazing off somewhere into space. Those who met Banach daily confirm the expression and his staring: for example, Andrzej Alexiewicz wrote that Banach's eyes always "made a big impression;" Kazimierz Kuratowski said that Banach had an "incredibly penetrating gaze," and a former female student from Lvov talked about "his fascinating eyes." In Ulam's memoirs, Banach is described as follows:

> He always looked very young to me. He was tall, blond-haired, blue-eyed, and rather strongly built. His way of speaking struck me as being very direct, forceful, and perhaps even overly coarse (that characteristic, as I realized later, was to an extent deliberately cultivated by him). His expression was usually cheerful albeit mixed with skepticism.

Most of his weaknesses seem to have been almost endearing. For example, Banach attended soccer games, considered lowbrow entertainment in those days; tennis and horse riding were the proper sports for members of the intelligentsia. He paid no attention to propriety and fashions, followed his intuition, and acted on the spur of the moment. This trait certainly served him well in his scientific endeavors. People delighted in telling stories about the amount of alcohol and coffee he consumed. However, he was never seen drunk in public. And, his habit never interfered with his work or with the fulfillment of his obligations towards his family, university, students, or collaborators.

The author believes that many damaging opinions, some of them circulating to this day, can be judged as unjustified – overdoing Banach's "evils with great gusto," so to speak. His half-sister A. Waksmundzka recollects:

> My brother drank coffee by the gallon and consumed a lot of alcohol. However, he had an incredibly strong head. Once he attended a mathematical meeting in Georgia. There is a tradition there that at a banquet one drinks to every participant's health in succession. Emptying a tall glass of strong Georgian vodka is the norm at each toast. Banach

outdrank all the seasoned local dinner participants, and stayed on his feet even after other mathematicians one by one disappeared under the table following yet another toast.

Of his humor, the main characteristic was a certain irony tinged with pessimism. He was the Dean of Faculty and had to participate in all kinds of meetings. He tried as much as he could to avoid those boring occasions and used to confide in friends, "I know where I'm not going to be."

It is evident that for the frequenters of the Scottish Café, mathematics was not hard labor but pleasure. The old and the young, professors and assistants, and even students all had great fun doing mathematics.

It was said about Banach that he could work in his university office as well as in the waiting hall of a railway station. This disregard for the ivory tower was sometimes used against him. All the reports agree that Banach was a workaholic. Andrzej Turowicz notes:

> That is true, but his hard-working habits were unlike those of other people. A scholar usually needs peace and quiet to do his work. Not so for Banach. He would spend most of his days in cafés, not only in the company of others but also by himself. He liked the noise and the music. They did not prevent him from concentrating and thinking. There were cases when, after the cafés closed for the night, he would walk over to the railroad station where the cafeteria was open around the clock. There, over a glass of beer, he would think about his problems. To be honest, I have to add that he was not always able to master every problem quickly. He told me once that one question occupied him for several months, many hours daily. But in less complex cases his theorems squirted out, as if from a fountain. It was truly amazing and extraordinary.
>
> He worked more than twelve hours a day, and the fact that it was mostly in cafés is beside the point. He had no other special interests outside of mathematics. Some soccer games, some chess, a game of cards now and then, a trip to the mountains to Zakopane, or to the mountain village of Skole near the town of Stryj – these were his only moments of rest and relaxation.

Numerous reports confirm that he had little interest in anything besides mathematics. People referred to him as a one-sided genius. Some of his contemporaries claim that political affairs were of no interest to him.[1] During the war, his views were far from "realistic,"[2] as Steinhaus wrote:

> Banach was a mathematician first. Political affairs were of little interest to him, although he did have an acute understanding of each concrete situation in which he happened to find himself. The beauty of nature made no impression on him, and art, literature, and theater were just secondary entertainment, which at most – and very rarely – would serve to fill short breaks in his work. However, he cherished good harmonious company at the table.... He had no illusions and understood perfectly well that only two percent of the total population is actually able to grasp mathematics. Once he told me :
> – Know what, brother? I tell you that studying the humanities in high school is more important than mathematics – mathematics is too sharp an instrument, no good for kids.[3]

Banach looked like anything but a scholar. Steinhaus, who in some way knew Banach best, wrote the following in the fourth volume of *Wiadomości Matematyczne*:

> Anybody who imagined Banach as a dreamer, abnegator, apostle, or an ascetic would be in error. He was a realist, who even physically did not look like a candidate for sainthood.... Even as recently as 25 years ago there existed an ideal of a Polish scholar, formed not so much on the basis of real observations, as from the spiritual needs of the era which were so well expressed in Stefan Żeromski's [1864–1925, a famous Polish novelist] writings. A scholar was one who worked in the service of a not-so-well-defined society, removed from all the worldly temptations, the obvious futility of his work already forgiven. A scholar would not care that in other countries scientists were not measured by the degree of their personal sacrifices but by what they had contributed permanently to science. The Polish intelligentsia, even between the two World Wars, was still under the spell of this martyrdom ideal, but Banach was never influenced by it. He was healthy and strong, a realist to the point of being almost a cynic; but he gave Polish science, and in particular mathematics, more than anybody else. No one else contributed more to dispelling the damaging perception that in scientific competition a lack of genius (or even just the lack of talent) can be compensated for by other virtues. He emphasized his highlander background and had a slightly condescending attitude towards the type of intellectually educated person "without portfolio."

* * *

Banach spent much of his time with undergraduate students, and in writing elementary and academic textbooks. He was a department head, and a dean, who had responsibility for both the prosaic problems of undergraduates and the guidance of talented and gifted students. Kazimierz Szałajko, now a retired docent at the Silesian Polytechnic, who was Banach's student and later assistant, describes Banach as a professor and undergraduate advisor:

> Banach was, in general, very accessible: at any time one could walk into his office in the old building of the Jan Kazimierz University. During lectures students did not ask questions of professors, it just wasn't done. Students were shy, and there existed a considerable distance between the faculty and the student body. I remember that Banach's oral examinations were conducted in a peaceful and encouraging atmosphere. I proved a certain theorem and introduced a continuity assumption. Banach quietly explained to me: You do not know it yet, but there exists a generalization of this theorem which does not require this assumption.
>
> Later, at the seminar, Banach made an effort to get students interested in solving mathematical problems; he provided topics to choose from, and when we looked for sources on the subject he would send us to Auerbach – an expert on mathematical literature. In the 1930s, Banach was a supervisor of the student Mathematical-Physical Circle at the Jan Kazimierz University. Supervisors of scientific student circles were appointed by the Faculty Senate for a period of three years. The Circle organized, among other things, tourist excursions and student dances. Banach often took part in these events, although there was no pressure or requirement for him to do so.

Józef Jarymowicz, who majored in mathematics at the Jan Kazimierz University, says that "lucky was the class to which Banach was assigned as an advisor," and continues:

> Banach had the gift of crystal clear exposition of mathematical knowledge. At lectures, he spoke in a slightly hushed voice, but the room was quiet: everybody was entranced by what he was saying. He employed very simple language. Listening to his lectures we had an impression that he was telling us: Look how simple and easy this is. After his lectures there was no need to further cram the material. Leaving the classroom one understood the particular problems of a given subject. And for a mathematician, gaining understanding of notions, theorems,

or a theory is absolutely fundamental. As a rule, he was 10 to 15 minutes late for class, but in the remaining half an hour he would transmit to us more knowledge than could be delivered in several hours of somebody else's lectures. Compared to Banach, all the other instructors were very uptight in their lectures, and transmitted the material somewhat nervously. They couldn't do it the way he did it, simply and sort of off-the-cuff.

Another former student, Helena Opolska, remembers:

> We used to invite Banach to student balls, which were held in the men's dormitory on Łoziński Street, sometimes in the elegant Hotel George. One of the organizers would go over to Banach's apartment on Supiński Street and asked him and his wife to join us. At dances Banach used to dance with all the female students. He was a "professor-charming," very direct in his contacts with students. A superb dancer, he was also superbly demanding as an examiner. Once we organized an all-night ball at the Hotel George. We were still having a great time when, at six o'clock in the morning, the orchestra started packing up their instruments, indicating that the ball was over. Everybody felt sorry about it because the ambiance was magical. At that point Banach intervened. He approached the orchestra: "Let the youth enjoy it some more."
>
> He paid them for additional playing time and, turning to us, said, "Go ahead and dance for another hour."

In his 1992 inaugural address at the Silesian Polytechnic in Gliwice, Szałajko also reminisced that Banach had the reputation of a superb, albeit unorthodox teacher. In the fall quarter, his lectures did not start until mid-October although the official first day of classes was October 1. He used to spend the summer months working intensely in the quiet and cool surroundings of his office, located in the thick-walled ex-monastery building. In September he finally left for his vacation with his family, either in the Tatra Mountains resort of Zakopane or in the resort village of Skole on the Opor River, a tributary of the Dniester. The date of the first day of classes would be announced only after Banach's return from vacation. All the other professors followed the official university calendar and needed no special announcements.

Szałajko also clearly remembers that students never minded the irregularities in Banach's lectures, but rather recalled their clarity of thought, fluidity of delivery, and rigor, which made up for all departures from the norm. Sometimes Banach's lectures were canceled even in the middle of the academic year. Apparently, this was not an infrequent occurrence. The routine of such a cancellation involved a morning phone call from Mrs. Banach to the teaching assistant on duty (Szałajko in this case) with the concise message: "My husband does not feel well. Please cancel his class." Despite all of this, the syllabus of his courses was always covered in full by the end of the academic year.

It is interesting to note the difference of opinion between Ulam and Steinhaus concerning lectures Banach gave at the University and the Polytechnic. Ulam writes:

> I attended just a few of Banach's lectures and, by and large, they were not well prepared. Mistakes and omissions occurred. It was most fascinating to watch him at the blackboard, as he struggled, analyzed, and invariably conquered upcoming problems. I had considered this type of lecture more interesting than smooth expositions, during which my floating-in-space concentration came to earth only when I realized that the teacher was in trouble.
>
> From my junior year onwards, decisions on what mathematical classes I was to take originated in conversations with Mazur and Banach. According to Banach, my own research contributions were characterized by a certain peculiar strangeness in formulating problems and sketching possibilities of their proofs. He told me once, several years later, that he was surprised how often these strange suggestions proved useful in reality. Such a judgment, expressed in the words of a great master, was probably the greatest complement a twenty-year-old young man could encounter.

On the other hand, Steinhaus, from his perspective as Banach's discoverer and mentor, says something seemingly opposite:

> Banach... was a superb lecturer, he was never lost in details and never covered the blackboard with numerous and complicated formulas. He did not care about perfection of the verbal delivery; all humanistic polish was foreign to him, and for all of his life he preserved some of the characteristics, in speech and behavior, of a Cracow street urchin. The formulation of thoughts on paper was for him a painful process. He would write his manuscripts on loose sheets torn out of notebooks; if part of the text had to be changed he would cut it out, glue a clean

sheet of paper underneath, and fill in the new version. Were it not for help from friends and assistants, Banach's first papers would have never reached the print shop. He almost never wrote any letters....

He was also benevolent, even almost fatherly, towards his assistants. He behaved as their partner, devoid of condescension, paternalism, or superior feelings. Szałajko remembers Banach as a hard-working man, generous with his knowledge, stubborn, solid, with a sense of solidarity, reaching the goal by the shortest, and therefore the simplest, route:

> I remember the time when I was helping Banach to prepare a school textbook. Before the war, textbooks were published by major publishers. In Lvov, that role was fulfilled by the Książnica-Atlas Publishers. These people turned a tidy profit on textbooks but also were exposed to some risks. They would sign a textbook contract with an author, the book would be delivered, and yet the ministry of education could refuse to approve the textbook for official use. The publisher would thereby incur a loss.
>
> In 1936, Banach summoned me to his office and said that there was work available, and that he would like to ask me for help in preparation of a textbook. I accepted. For me it was a great honor to be able to collaborate with Banach. I knew that the textbook was supposed to be ready in a few days and that at that point Banach had just a few things jotted down on several loose sheets of paper. We agreed to meet again in a few days. I knew that he was working on the book's concept; the light in his office was on until late at night. It was Thursday when I returned and the work began in earnest. We worked frantically for four days. Besides me, Auerbach was also involved. Banach dealt with the theoretical part. Auerbach prepared problems and exercises, and also made a clean copy. We took only short breaks to walk over to the Scottish Café to reinforce ourselves with strong coffee and a snack. My eyelids felt as if they weighed a ton from weariness as we talked about what else was left to be done. I would take from them the handwritten pages of the manuscript, number the pages, proofread them, and dictate the text to the typist.
>
> The publisher left at our disposal a veritable stable of his employees. Besides the typist and printers, a special courier was waiting for the trial printout of the textbook to take it to Warsaw, so that it would be at the Ministry on Monday morning. The last pages were created on Sunday.
>
> I was back in Banach's office on Monday. He tore off a piece of scrap paper and wrote on it a couple of sentences instructing the publisher to

pay me 450 złotys. For me this was an enormous sum – my monthly assistant's stipend was only 160 złotys.

Banach wrote a number of elementary mathematics textbooks, some of them jointly with Sierpiński and Stożek. His spending habits were never under control and in this way he could supplement his income and pay off his debts.

* * *

We said earlier that Banach lived at peace with himself. He was cheerful, direct in his manner. Kuratowski noted how extremely communicative he was, and that because of this gift, Banach had a great influence on his numerous disciples and the Lvov community. He was modest but, at the same time, he knew his worth. Once he told his half-sister Antonina: "If you travel abroad and find yourself in need, just find some mathematicians and tell them that you are Banach's sister – you'll find the doors open everywhere." However, he never used his position and prestige for personal gain.

Bohdan Miś, a popularizer of mathematics and a journalist to whom Banach's son later made the original *Scottish Book* available, relates the following perhaps partly apocryphal but revealing story, which became in Poland a part of the Scottish Café legend:

> ...Von Neumann (1903–1957), an American mathematician of Hungarian descent, called by some "the Gauss of the 20th Century," visited Poland three times between the wars. Each time, on personal instructions from Norbert Wiener, father of cybernetics, he tried to talk Stefan Banach into emigrating to the United States; his last visit to Lvov took place in 1937. Responding to the latest job offer Banach asked:
> "And how much is Professor Wiener willing to pay?"
> "We anticipated this question," responded the confident American reaching into his pocket; "here is a check signed by Professor Wiener on which he entered only the numeral 1. Please add to it as many zeros as you deem fit!"
> Banach contemplated the offer for a moment and responded: "This sum is too small to leave Poland."
> And he stayed, as the clouds of war were beginning to gather over Europe. Even at the last moment before the war, when Ulam proposed an escape plan, the scientist refused to leave. That hard, coldly calculating

man was a great Polish patriot, and he felt a greater attachment to Poland than anything else. . . .

In 1948 Hugo Steinhaus wrote about him in *Colloquium Mathematicum*:

> . . . Banach was not a mathematician of finesse, he was a mathematician of power. . . . Inside, he combined a spark of genius with that amazing inner imperative, which incessantly whispered to him, as in Verlaine's verse, "There is only one thing: that intense glory of the craft" – and mathematicians know well that their craft depends on the same mystery as the craft of poets.

CHAPTER VII

The Last Years

ON SEPTEMBER 17, 1939, after the Ribbentrop-Molotov Pact, the Soviet Union entered the war, and Soviet troops soon occupied Lvov. The Soviet "Anschluss" of Lvov was followed by massive deportations and persecution of the Polish population.

Inquiries about Banach started almost immediately. Soviet mathematicians highly valued Banach's contributions to science and showed their appreciation for him on numerous occasions. Thus Banach maintained good relations with the Russians and the Ukrainians. He was offered by the new administration a continuing position at the Jan Kazimierz University – now renamed Ivan Franko University in honor of a Ukrainian poet – at the same time that dozens of new Ukrainian faculty members were being brought from Kiev and Kharkov. Banach was made Dean of the Physical-Mathematical Faculty and Head of the Department of Mathematical Analysis. Professors Orlicz, Saks, and Knaster were employed in his department. He was invited and traveled to Moscow. His *Theory of Linear Operations* was immediately translated into Ukrainian. Soviet mathematicians Anisim Bermant, Lazar Lusternik, and a few others traveled from Moscow to Lvov. When Bermant was asked what were the most important mathematical centers in the USSR he responded: Moscow in the first place. But he had doubts about the second place: Leningrad or Lvov.

During the first days of World War II, Banach's father Stefan Greczek and his half-sister Antonina arrived in Lvov from Cracow, escaping before the advancing German armies. Only then, recalls Antonina, did they become friends and emotionally close, the way siblings should be:

> My brother, more than 20 years my senior, bestowed on me his special affection and trust. He confided in me about his intimate and personal affairs, for example concerning relations with his wife Łucja.

Banach was happy about the arrival of his relatives. He got satisfaction from the fact that now it was he who could provide for his father and half-sister, and share with them the respect and social status that came as a result of his own work and scientific achievements.

At that time, Banach's lifestyle was not much different from what it was before the war. He lectured, continued his research, was involved in civic and administrative activities, visited the Scottish Café, and kept writing school textbooks.

The *Red Banner*, the official Soviet Lvov newspaper of the period, reported at the end of 1940 in its inimitable prose:

> ... Within the walls of the Lvov University, the forge of knowledge in the Western territories of our Republic, there work a number of distinguished scholars who – under most favorable conditions created by the Soviet power – pursue their broad programs. At the Lvov University there exist 65 scientific departments and divisions, and the faculty thereof is currently engaged in work on 477 scientific projects.

Further, a piece of information about Banach followed:

> ... A textbook in the area of theoretical mechanics is being written by Professors Banach and Schauder.

Two months earlier, the *Red Banner* provided the public with more detailed information about Banach's activities:

> ... The professors of Franko University prepared for publication a number of scientific papers which could not be published until now, and

completed many other scientific projects. They will be published in scientific bulletins of various faculties, as is the custom at other Soviet universities and institutes.... The collection of papers of the Physico-Mathematical faculty contains 23 memoirs. Soon, the university will also restart the publication – in three foreign languages, French, English, and German – of the great international mathematical journal *Studia Mathematica*, under the editorship of Professor Banach. Eighteen papers of distinguished scholars from Lvov, Leningrad, Moscow, Kharkov, and abroad will appear in the new issue. The hundreds of letters that flow to the Faculty from all corners of the world are a vivid testimonial to the great interest that the scientific world has in the work of our mathematicians and physicists....

The same newspaper informs us that in 1940 Banach was elected to the Lvov City Council. It is known that he used his influence there to help people in the persecuted Polish community. His very close contacts with Soviet scholars continued. Pavel Aleksandrov, a distinguished Soviet topologist who wielded major influence in the Russian mathematical enterprise, later wrote in a letter to Kuratowski: "In those years, we also became well acquainted with Mr. Banach, whom I've seen in Lvov and who visited us in Moscow many times where I had the honor to host him at my home on several occasions." Sergei Sobolev, a major Russian analyst, who for the first time visited Lvov in 1940 together with Aleksandrov, has similarly warm recollections of Banach. Alexandrov's and Sobolev's entries in the *Scottish Book* are from that period. Also at that time, Banach was elected Corresponding Member of the Ukrainian Academy of Sciences in Kiev.

* * *

The abrogation of the Ribbentrop-Molotov treaty and the German invasion of the Soviet Union caught Banach in Kiev, where he was attending a mathematical meeting. He immediately boarded a train – the last that departed from Kiev to Lvov – and, disregarding personal danger, returned to his wife and son. When in June of 1941 Hitler's troops entered Lvov, the Banachs found themselves in a very difficult situation. Banach worried that the Germans would ask him to account for his cozy relationships with the Soviets. Also, the very fact that he was a member of the Polish intellectual elite put him in danger. He was arrested by the Gestapo on suspicion of trafficking in German currency but released after a few weeks. Some of his tenants were apparently involved in that activity. While in jail he even managed to prove a couple of theorems.

Józef Sieradzki, a Lvov resident during the Soviet occupation, describes the atmosphere reigning in the city:

> I can state without exaggeration that we died at every sunrise. Sleep was the only time when we could relax. Even those rare hours, when each of us – dead tired and irritated by the disgusting labor we were forced to perform – escaped to the distant and vanished world of past scholarly pursuits, ceased to exist with the arrival of the Germans.

It was a tragic time. By Himmler's order issued within the so-called *Ausserordentliche Befriedungsaktion* (Extraordinary Pacification Action), the Lvov elite were to be entirely eliminated. In Cracow, the professors had been summarily arrested and deported to concentration camps two years earlier. The liquidation of the Lvov intelligentsia had been already planned in 1939. Two lists of people to be executed had been prepared. The first contained names of scientists employed by the Polytechnic and the Merchant School; the second, names of university professors. The preparations were meticulous, and even the place of executions had been selected in advance. The murderous plans were carried out in secrecy, as opposed to the more public actions in Cracow that caused major reverberations in Western public opinion. The special liquidation commando appeared, entered into action just behind the advancing front-line troops, did his work, and then disappeared. The action appeared to have had no official Wehrmacht sanction.

During the night of July 3, 1941, 40 Polish scholars, professors, writers, and other distinguished representatives of the Lvov intelligentsia perished at the hands of the Nazis and the Ukrainian nationalists from the S.S. "Nachtigall" battalion. Included in the group were writer and journalist Tadeusz Boy-Żeleński, and Banach's friends and colleagues Włodzimierz Stożek, surgeon Tadeusz Ostrowski, Antoni Łomnicki, and Stanisław Ruziewicz.

The executions were only the beginning of a broader action aimed at the destruction of Polish intellectual life in Lvov. After the war, Józef Sieradzki tried to find some rationale for the massacre:

... The German crimes had their theory and their methodology. A study of the origins and psychology of these crimes leads into a thicket of convoluted and murky questions, but the goal of the Lvov action, carried out at the very moment when Himmler, the chief commander of the Teutonic Order, was visiting in the city, is quite obvious. The action was executed with lightning speed, without any pretense of due legal process; even the most basic court formalities routinely accorded to repeated murderers were skipped. The action of destruction of the scholars was completed so fast that even some of the German authorities were caught unaware. A few days after the massacre, emissaries of oil companies from Berlin tried to approach the just-murdered Professor Pilat, an internationally known expert on oil and gas technology.

There is proof that the henchmen had a prepared list of their victims. This list must have been prepared in 1938–39 since Gestapo men were looking for two scholars who died between 1939 and 1941: a dermatologist, Professor Leszczyński, and an ophthalmologist, Professor Bednarski. In both cases, their widows were required to shows certificates of their husbands' deaths. The flowers of science and literature were cut down. The luminaries of knowledge. Almost every name from that list had appeared on dozens of scientific papers, signified a discovery, defined a theory, was an ornament of academies or learned societies. They were among the world's top intellectuals, decorated by the laurels of international awards, distinctions, and medals. The eternal Teutons destroyed the Lvov scholars, the same way they exterminated the Kashubs a few centuries earlier in Gdansk, just because they were Slavs and Poles and not Germans.

Banach did survive this pogrom of the Lvov intelligentsia. However, the conditions under which he had to live during the Nazi occupation were very harsh.

Jadwiga Hallaunbrenner, a Lvov mathematician, whose husband Michał was a physicist, remembers:

> Banach visited us regularly. My husband, thanks to the intervention of Kazimierz Ajdukiewicz, got a job at Viehverband, the butchering and meat packing company which bought cattle and produced second-class sausage. He brought home a kilogram of meat every week – it was the worst kind of sausage and tripe. I would pull down the kitchen window blinds and cook a big pot of beans with meat. In the evening Stefan Banach, Tadeusz Riedl, and the Polish literature professor Kazimierz Brańczyk, and perhaps some others whom I do not remember, would

show up. Banach was exhausted, starved, and wasted away, although before the war he was a square-shouldered and strong man. After the men finished eating, a conversation was begun: about the situation at the front, when the allied landing was to be expected, what was going on at the Eastern front, what was up in Poland, in Lvov, what were the chances of survival, etc.

Banach's pitiful physical condition was a result not only of the general famine – at the time, it was a great success to secure even a small chunk of horse meat – but also of his job as a feeder of lice in the Rudolf Weigl Bacteriological Institute. He started working there in the Fall of 1941. The job had its good side. An identity card from the Weigl Institute permitted one to survive the occupation in relative security; it was sufficient to get through cordons of gendarmes randomly rounding up the Polish population. It created a sort of safe official status. Stanisław Ruziewicz's son Zdzisław remembers this period:

> ... The Institute employed everybody who had any direct or indirect contact with scholarly work, that is, a majority of the Lvov intelligentsia. I also worked at Weigl's Institute. We fed lice sitting at a long wooden table. I remember that almost instantly social groups began to form. There was a table where only humanities professors were feeders. Feeders who were natural scientists occupied another table. Banach and Knaster sat at the same table, and I had the impression that they were engrossed in mathematical conversations there. I know that during the war Banach gave private mathematics lessons to Tadeusz Riedl's son.

Rudolf Weigl discovered an active vaccine against typhoid fever and his Institute conducted research on the disease. It was then a very dangerous illness. In 1942, 90,000 inhabitants of the General Government – the central part of Poland under German administration – succumbed to it, and by a decision of the commander-in-chief of the German land forces (OKH), a prophylactic method of fighting typhoid fever was adopted. The anti-typhoid vaccine was manufactured, among other places, on a huge scale at the Behring company in Lvov.

Banach worked as a feeder of lice throughout the remainder of the Nazi occupation of Lvov, that is, until July 1944.

* * *

On July 27, 1944, Soviet troops reentered Lvov. Immediately thereafter, the Banachs moved to the Riedls' house on Dwernickiego Street.
Banach's friend Dr. Tadeusz Riedl was a lawyer, but also owned a plant-
seeds distribution business.

Banach soon renewed his Russian contacts and was drawn into the
activities of the All-Slavic Antifascist Committee – a Soviet-sponsored
organization. The committee, as reported in the Soviet press, "... was
created on August 11, 1941, in a public meeting attended by representatives of all the Slavic nations. It achieved great success in advocating the
unity of Slavicdom in its struggle against German Fascism, and inform-
ing world public opinion about Nazi crimes in the occupied territories of
the USSR, Poland, Yugoslavia, and other countries." At that time, Sergei
Sobolev met Banach often at the *Uzkov*, an exclusive resort hotel of the
Soviet Academy of Sciences located some 10 miles outside Moscow:

> Despite heavy traces of the war years under German occupation, and
> despite the grave illness that was undercutting his strength, Banach's
> eyes were still lively. He remained the same sociable, cheerful, and
> extraordinarily well-meaning and charming Stefan Banach whom I had
> seen in Lvov before the war. That is how he remains in my memory:
> with a great sense of humor, an energetic human being, a beautiful soul,
> and great talent.

An offer of a chair at the Jagiellonian University arrived. Banach
accepted it with gratitude and joy. It also served as a pretense for his formal repatriation process, from the now Soviet-Ukrainian Lvov to Poland.
The Editorial Committee of the reborn *Mathematical Monographs*, now
headed by Kuratowski, with Banach as a member, resumed its activities. The war interruption notwithstanding, Banach was still the sitting
President of the Polish Mathematical Society. He was also offered the
position of Minister of Education in the new – Soviet imposed – Polish
Government. However, despite physicians' efforts, the spreading lung
and bronchial cancer could not be stopped. One of his female students
recalls:

> I and my husband visited Banach three days before his death. He
> was sitting in his bathrobe, almost entirely silent; he just uttered a few

words in a gravelly voice. His complexion was ashen and there were crimson spots on his face.

Stefan Banach died on August 31, 1945, in Lvov in the apartment at 12 Dwernickiego Street. He was just 53 year old and still bursting with plans for the future. The news about the great success in the U.S. of his former students and collaborators had just begun to filter in from behind the Western front. The newspapers of the period showed the depth of mourning in the Lvov scientific community upon his death. His funeral was attended by hundreds of people. The sidewalks of St. Nicolaus Street, which he had walked so often in the past, were lined with female students of the former Jan Kazimierz University, each one holding a bunch of flowers. Banach's remains were laid to rest in the family crypt of the Riedls. The tomb is located at the Łyczakowski cemetery, by the no-longer-used gate, formerly opening on St. Peter Street. In 1990 his ashes were moved to the Crypt of the Distingushed at the Na Skałce Church in Cracow.

Five of Banach's papers appeared posthumously. Also a university textbook, *Introduction to the Theory of Real Functions*, appeared in 1951 as volume 17 of the Mathematical Monographs series. He had worked on them during the war years.

CHAPTER VIII

In the Eyes of Friends and Followers

"A mathematician is a person who can find analogies between theorems; a better mathematician is one who can see analogies between proofs and the best mathematician can notice analogies between theories; and one can imagine that the ultimate mathematician is one who can see analogies between analogies,"

said Stefan Banach in one of his speeches. On another occasion he also wrote:

Mathematics is the most beautiful and most powerful creation of the human spirit. Mathematics is as old as Man.[1]

In 1960, on the 15th anniversary of Banach's death, the Institute of Mathematics of the Polish Academy of Sciences organized a conference to commemorate and evaluate his influence as a mathematician and his impact as a person. Several of Banach's contemporaries contributed to the Proceedings which were published in 1961 in the fourth volume of *Wiadomości Matematyczne*, the bulletin of the Polish Mathematical Society. The remainder of this chapter contains excerpts from that volume.

Banach's "discoverer," *Hugo Steinhaus*, wrote about Banach's approach to mathematics in this way:

> He was not enamored with logical dissections, although he understood them very well; he was not attracted to practical applications of mathematics either, although he certainly could do them successfully – after all, one year after his doctorate he taught mechanics at the Polytechnic. He used to say that mathematics identifies itself by a specific beauty and will never lend itself to a reduction to a rigid deductive system. Sooner or later it will outgrow any formal framework imposed on it and will create new principles. The value of mathematical theories was decisive for him, but it was an inner value [that was important], not a utilitarian one. His foreign competitors in the theory of linear operations either considered spaces that were too general and as a result obtained mostly banal results, or made too many assumptions on these spaces, which narrowed the range of applications to a few artificial examples. Banach's genius was to find the golden mean. This ability to hit the bull's-eye certifies Banach as a pure-bred mathematician. . . .

Sergey Sobolev, a Soviet mathematician who decisively contributed to the creation of the modern theory of distributions, a branch of functional analysis of utmost importance in applications of the latter to mathematical physics and partial differential equations, wrote:

> This outstanding and serious mathematician, one of the creators of functional analysis, the most important contemporary direction in mathematics, left humanity, through his numerous papers, through the creation of his own mathematical school, through his disciples and followers who can be found all over the globe, a series of momentous results – magnificent achievements of the human genius.
>
> The first half of the 20th century was an epoch of unusual discoveries in physics and mathematics. As a consequence of the revolution in physics caused by discoveries of relativity theory and quantum theory, the face of contemporary science was completely changed, and

the world outlook of scientists has been altered. Classical perceptions of space, time, and physical quantities disappeared altogether. In the modern approach, the physical quantities are operators – a notion unknown in the 19th century. The whole complex of ideas in modern physics, and the ensemble of its basic concepts, was created thanks to the achievements of new mathematics.

The revolution in mathematics prepared the ground for new physics, running its course in parallel with the revolution in physics, and perhaps even ahead of it. Old notions were neither rejected nor overturned. This would not have been in the mathematical character. But in mathematics, as in physics, it was unexpectedly discovered that there exists a new, immeasurable world, a new universe where previously existing mathematical results can be seen in a different light. By now there remains not a single branch of mathematics where the influence of ideas of functional analysis, personifying the hot breath of modern times, cannot be felt.

At the cradle of functional analysis, standing out among its creators, was the great Polish scholar Stefan Banach. . . . I would like to characterize, in a few sentences, the most important of his scientific achievements, pause to contemplate the significance of his papers for the development of classical analysis, numerical methods, the theory of partial differential equations, integral equations, and related areas.

First of all one should mention, of course, the theory of normed complete spaces, which today are called Banach spaces. That theory contains many achievements of the first magnitude which were obtained by Banach, such as the theorem on the extensions of linear functionals, the theorem on the boundedness of the inverse operator, the theorem on weak convergence of operators, the theorem about the possibility of the embedding of a separable space in the space C of continuous functions, and many others.

Contemporary papers in the area of numerical methods are no longer a collection of recipes to solve this or that practical problem, as was the case before Banach. In this field the topic of studies itself has changed. Now, it is always an investigation of concrete methods of constructions of nets in compact subsets of Banach spaces. Due to these general ideas, the goal and direction of this discipline has been altered, and its main problems and general methods have become clear. Without Banach spaces, the modern theory of numerical methods could not exist.

Also, there are no papers in the modern theory of partial differential equations, where, at the most fundamental level, one could not find the concept of a solution as an element of a certain functional Banach space.

These new ideas were devised while Banach was still alive, by himself and by his closest collaborators and students.

The classical notions of existence and uniqueness of solutions were later augmented by an extremely important notion of the well-posed problem, i.e., a problem with a continuous dependence of solutions on the boundary values and other conditions. But the continuity can almost always be expressed in terms of a Banach space: small changes of conditions in the sense of a Banach norm correspond to small changes of solutions in terms of another Banach norm.

The influence of functional analysis is not limited only to posing problems and formulating basic notions. The fresh contemplation of the contents of the existing methods for solving certain problems led to the expansion of the areas of its applicability and, in many cases, led to the creation of fundamentally new methods.

Applications of functional analysis in mathematical physics, based on the theory of contracting operators, on the theory of inverse operators, and on extensions of functional spaces, became almost ever-present, pushing out in the process the classical, often algorithmic, methods based on ideas from function theory.

Stefan Banach's *Theory of Linear Operations*, as was the case with other classic treatises, became the property of the wide world of mathematicians. The notions introduced in it, particular theorems and whole theories, firmly established themselves in the conscience of each of us.

The Polish nation, having given the world the gift of people like Frédéric Chopin, Adam Mickiewicz, and Maria Skłodowska-Curie, who entered permanently in the history of human culture, is justified

in being proud of its distinguished son Stefan Banach, whose name
will be permanently connected with the development of mathematics
in the 20th century.

Stanisław Mazur, Banach's student, collaborator, and friend from the
Lvov times, wrote:

> The existence of functional analysis as an independent mathemat-
> ical discipline is indebted to the genius of Stefan Banach. He shaped
> its basic concepts, and the fundamental theorems were proved by
> him.
> The creation of functional analysis, as in the creation of any
> other new scientific discipline, was the last stage in a long historical
> process. Voluminous is the list of mathematicians whose research
> contributed to the formation of functional analysis. It includes such
> famous names as Vito Volterra, David Hilbert, Jacques Hadamard,
> Maurice Fréchet, and Frederic Riesz. But the year 1922, when Stefan
> Banach published in the Polish journal *Fundamenta Mathematicae*
> his doctoral dissertation *Sur les opérations dans les ensembles ab-
> straits et leur application aux équations intégrales*, is unquestionably
> a threshold date in the history of 20th-century mathematics.
> That memoir of a few dozen pages put the foundations of func-
> tional analysis in their final form. As a result of the efforts of Ste-
> fan Banach and others, a new mathematical discipline was created
> which has a capital significance for the further development not only
> of mathematics itself but also for natural sciences and, in particular,
> physics.
> Functional analysis replaced the notion of a number, which was
> fundamental for mathematical analysis, by a more general notion
> that today, in thousands of mathematical memoirs, is called by the
> term "the point of a Banach space." A generalization of mathemati-
> cal analysis obtained in this fashion allowed one to treat in a simple
> and unified way seemingly different questions of mathematical anal-
> ysis, and to solve many of the problems with which mathematicians
> vainly struggled before. It also significantly increased the help that

mathematics could offer to the natural sciences, particularly physics. However, the importance of mathematics in science relies not only on the fact that it permits us to derive conclusions concerning the dynamics of phenomena on the basis of verified or historical regularities, but also on the fact that, as it develops, it creates new notions. As a result, it becomes possible to translate into mathematical language new classes of phenomena, which consequently can lead to their better understanding. Among other things, functional analysis created an appropriate apparatus of concepts that permitted construction of models for phenomena that are currently a subject of study in modern physics.

During the less than 40 years that have passed since the appearance of Stefan Banach's doctoral dissertation, functional analysis has grown to a mighty branch of mathematics, which attracts the attention of more and more mathematicians all over the world. Thanks to studies by Polish mathematicians, as well as research conducted in the great centers of functional analysis in the Soviet Union, United States, and France, Stefan Banach's ideas have been significantly broadened. All the up-to-date development of functional analysis shows that Banach's edifice has proved to be of great permanent value for science. Functional analysis itself is a magnificent monument to its creator.

Bela Szökefalvi-Nagy, a Hungarian mathematician and coauthor of one of the first monographs on functional analysis, added:

> Paying respects on behalf of Hungarian mathematicians, who hold the memory of Stefan Banach in the highest esteem, I cannot omit close ties connecting the works of Stefan Banach with the works of our great teacher Frederic Riesz, who passed away four years ago. As a matter of fact, the classic Riesz papers on functional spaces, on linear operators, and transformations in those spaces inspired to a high degree the far-ranging investigations of Banach and his collaborators. The results of these studies were presented in Banach's magnificent oeuvre *Theory of Linear Operations*. Without

any doubt it is one of those monographs that exerted the greatest influence on the development of contemporary mathematics. Although the theory developed in that book was preceded by the appearance of E.H. Moore's *General Analysis* and prepared by research on abstract spaces by M. Fréchet and others, and although the theory was able to utilize methods previously developed for more particular situations, and although the authors thereof, for example Frederic Riesz, predicted and explicitly stated their general usefulness, the theory was created almost in its entirety by Banach and his collaborators. Frederic Riesz always spoke about the value of that book with the highest respect, and expressed this opinion in his analysis of the *Theory of Linear Operations* written for the Szeged journal *Acta Scientiarum Mathematicarum.*

The theory developed in Banach's treatise encompasses in its methods a great variety of problems: first and foremost the problems of existence for differential and integral equations, and even more generally, for linear functional equations, then systems of linear equations with infinitely many variables, Fourier series, summation of divergent series, and finally functions without derivatives. Among the methods used one can find methods extremely ingenious and deep, as well as others, perhaps less effective, but amazingly simple instead. Riesz, in his analysis of Banach's book, especially praised the Banach-Steinhaus theorem on uniform boundedness, which is a generalization of Osgood's theorem for usual continuous functions and continuous operations in general linear spaces. This theorem implies, among other things, the results of Alfred Haar and Henri Lebesgue on the convergence of singular integrals, Hellinger and Toeplitz' theorems asserting that the sequence of quadratic reducts of infinitely many variables can converge in all points of a Hilbert space only when the form is bounded, and theorems about the existence of various categories of functions without derivatives. The theorem contains results belonging to different theories, and one has to remember that the discoveries of these results were major events in their own right.

The theorem about uniform boundedness, the Hahn-Banach theorem on extensions of linear operations, and Banach's theorem (often called "the closed graph theorem") represent three truly fundamental results of linear analysis. Messrs. Dunford and Schwartz correctly place these three "cornerstones" at the beginning of the recently published monograph on linear operators, a monumental book which makes a deep impression and gives a survey of the theory inaugurated by the papers of Frederic Riesz, Stefan Banach, and the Polish school, among others.

Marshal Harvey Stone, an American functional analyst who made major contributions to the spectral theory of operators, wrote:

> Stefan Banach's imprints on the mathematics of our century ensure for him a permanent place in the history of science. Both in view of his famous work, as well as of his contribution to the awakening of research interest and activities of mathematicians in his country, Poland, and also in other countries, he exerted a decisive influence on the development of contemporary functional analysis. Many among us, meeting here to honor this great Polish mathematician, are conscious of the influence of his ideas on our own investigations in the twenties and thirties. Others, whose own activity began a little later, remember him as a master in whose original treatise *Theory of Linear Operations* they sought knowledge and inspiration. The respect that we are now paying to Stefan Banach come both from our hearts and our minds. All of us who are fortunate to be present here and participate in this celebration to express our admiration for him, have colleagues in many countries who regret that they could not be here with us.
>
> Pausing for a while to contemplate our debt to Banach, we should devote some time to ponder the reasons which permitted his work to have made such a big and overwhelming impact on the development of functional analysis. Since mathematics becomes more and more complicated and detailed theories abound, we all feel a growing need to find a general perspective and a broad view which would lead us, by the most beneficial routes, to deep mysteries which we want to study.

If such an enlightenment is to be found at all, its most likely source will be an analysis of successes as well as failures of mathematicians whose work has already been completed so that we can still clearly see, in considerable detail, the mutual interaction of their works with the works of their contemporaries and direct followers.

Stefan Banach's example gives us many valuable clues. His work was based, as is the case of all of us, on the achievements of many of his illustrious predecessors, among whom we should mention Volterra, Fréchet, Hilbert, and Riesz. The main directions of functional analysis were already charted before Banach began his work. The role of fundamental linear-algebraic structures was clearly emphasized by Fredholm, Hilbert, and Riesz; the role of topological considerations was clearly understood – they were initiated by Fréchet; the generalization and the abstraction were clearly established as desirable and correct approaches to and goals for functional analysis (Riesz and E.H. Moore).

And yet, it seems that early Banach papers, created under conditions described by Professor Steinhaus, indicate that Banach's independence of spirit has to be valued very highly. These conditions could have prevented Banach from familiarizing himself with the achievements of his predecessors. During and immediately after World War I, Poland certainly was not the best place for a young mathematician to start a career – even for someone in personal circumstances much more comfortable than Banach's. On the other hand, the great enthusiasm for mathematics, which accompanied the creation of the magnificent mathematical school at the beginning of the twenties, provided Banach with an appropriate and extraordinarily stimulating atmosphere in which to develop his own ideas. Obviously, he benefited from the very lively interest in Poland in problems of set theoretic topology, which he exploited without forgetting about his own goals in analysis. Due to this fact, his achievements were characterized by a masterly utilization of topological methods to obtain deep theorems in functional analysis which escaped the notice of his predecessors and contemporaries. Banach presented his ideas in a mature and compact fashion in his famous monograph, emphasizing with an amazing clarity a subtle interplay between topological and algebraic considerations which helped create the truly fruitful abstract and general notions which contemporary functional analysis had to deal with. What made the influence of Banach's work so strong was the unification of several, discovered before him, results from analysis, which up to his time were disconnected and incomplete.

Although Banach saw the generalization as a goal in itself, we can see today that he actually did not pursue this direction as far as modern analysis requires. I'm sure this can be explained by Banach's acute intuition as to the best strategy that should have been followed in that period. Undoubtedly, Banach would understand and appreciate the current interest in topological groups and linear topological spaces, but at the time he was writing his book, the most interesting analytical problems were connected with metric spaces. Although the effort to generalize the theory to non-metric spaces was obviously desirable, it was undertaken only a few years after the appearance of his book. Most certainly, Banach felt that at the beginning one had to take care of the most important case, and use tools which were demonstrably ready for such a use at that very moment. Whether or not he really felt this way, we can agree *post factum* that he was right in pushing generalizations only to the point where deep results could already be obtained and where their direct application to solve the most interesting problems of the time was possible. To tell the truth, it seems to me only fair to mention that the work in the area of linear topological spaces and locally convex linear topological spaces completed at the end of the thirties bore fruit only after Laurent Schwartz formulated his theory of distributions, which showed new direction and supplied new incentives for studying those spaces.

Despite the intense studies of Banach spaces conducted since their introduction by Banach and, independently, by Norbert Wiener, their theory is still very far from being complete. It seems that not only many unsolved problems of that theory, some of them formulated by Banach himself, but also some of the most natural directions of future research, encounter very serious difficulties. To understand this, it is sufficient to realize how little of our knowledge of Hilbert spaces currently extends to Banach spaces, or more generally, to linear topological spaces. The specific linear metric structure of Hilbert spaces gives us a very strong foundation which is missing in other cases. Without that structure our efforts at obvious generalizations, for example in spectral theory, lead to at best partial results. There are many chapters of analysis which at present can be considered only in the framework of Hilbert spaces. It is worth emphasizing that this malady affects not only spectral theory but also other areas where more elementary considerations from functional analysis are applied. In particular, many contemporary achievements of the theory of partial differential equations rely on the ingenious applications of Hilbert space reasoning. It is a clear retreat from the earlier goal of utilization of Banach norms specifically designed for each problem. So we could claim that Banach's genius created for us as many problems as it solved.

Unquestionably, that was what Banach had to do. We who proceed along the road charted by him can only be grateful to him both for the light he shed on so many aspects of functional analysis as well as for the many problems he left for us to solve.

Stanisław Ulam, Banach's student and friend, published in the 1971 edition of *Wiadomości Matematyczne* his reflections in which he tried to place Banach's phenomenon, and in general phenomena of the Lvov mathematical school, in a broader context:

> The development of mathematics throughout human history has occurred under either the impact of certain communities or of some groups of people. The communities, both large and small, usually were created around one or sometimes several individuals, and sometimes were a result of the scholarly work of several people, more or less equally, who simultaneously worked and developed mathematical activity. Such a group would form more than just the commonality of specific interests; it would have a very concrete ambiance and character both in the selection of problems and in the thinking style.

At first sight it may seem strange, since mathematical discoveries, whether a new pregnant definition or a complicated proof solving a long-standing open problem, seem to be the result of completely individual efforts. They seem to be almost like musical compositions, and it is not easy to imagine how the latter could be written by more than one individual.

However, in a group of individual mathematicians, a choice of this or that problem and/or method is often repeated. This is a function of common interests. Such a choice is influenced by the interaction of questions and answers that certainly can be posed and solved by a single mathematician, but which emerge in a more natural fashion if they are the result of the work of several minds working together as a group. The great mathematical centers of the 19th century in Göttingen, Cambridge, Paris, and Russia exerted a special and definite influence on the development of mathematics in exactly this fashion.

A significant portion of Polish achievements during the twenty-year period between the World Wars contributed to the creation of the foundations for contemporary global mathematics. However, their influence was exerted not only on the subject matter but also on the tone of contemporary mathematical discourse.

Since Cantor's times, mathematics has been permeated more and more by the spirit of set theory. Recently, we witnessed a renaissance of interest in that theory and its unexpected progress. What I have in mind is not only set theory in its most abstract form, but also its direct applications: topology in its most general setting, most general algebraic ideas. The impulse and direction for all of that came from the Polish school.

Lvov mathematicians were responsible for a significant part of this contribution. Their studies were not solely concentrated on set theory, but also on a new framework for classical problems, which turned out to be functional analysis with its geometric and algebraic spirit. If one would like to give a very simplified description of the historical sources of this activity, it could be said that the

Polish school's work was based on Cantor's work, on the work of
logicians of the German school, and of the French mathematicians
Baire, Borel, Lebesgue, and others.

These studies, together with analytic problems formulated by
Hilbert and others in Germany, led to simple, general constructions
of infinite dimensional functional spaces. In parallel, and in a sense
independently, the American papers of E.H. Moore, O. Veblen, and
others, stimulated by general trends, led to a convergence of different
viewpoints and a unification of various mathematical intuitions.

The important feature of modern mathematics, which has been
fully implemented in Lvov, is a collaboration among different individuals, and even among completely different schools of mathematics. Against the grain of diversity and specialization, and even
hyperspecialization of mathematical research, the directions and research threads derived from different and independent sources often
have a lot in common.

I will not try to give a historical description and genesis, or even
a philosophical explanation, of this magnificent Lvov community. I
will only provide my personal impressions, both as a student and as
a participant, of the spirit and character of the group of faculty of the
University and Polytechnic in Lvov.

One has to remember that I'm writing these recollections about
the period between the two World Wars, after thirty years spent in
the United States. I have had only sporadic contact with Polish
mathematicians, with the exception of a short period just before
World War II when I visited Lvov during my summer vacations.

The kaleidoscope of types of Lvov mathematicians presented a
great variety of mathematical individuals, not only in terms of their
interests and education, but also in their mathematical intuitions and
customs. The areas with the set theoretic flavor, foundations of
set theory, set theoretic topology, and then – under the influence
of Banach and Steinhaus – functional analysis with applications to
classical analysis, were the main engines of original research in Lvov.

Schauder, who was a Docent at the University, also worked on
partial differential equations. His methods and results have become

classic by now and are one of the most powerful tools for proving existence theorems.

Banach, Mazur, and Schauder were the creators of – so popular today – the method of approaching analysis problems via geometric techniques of functional spaces.

If I cared to define the single most prominent characteristic feature of the school, I would mention its interest in the foundations of various theories. What I mean by this is that if one imagines mathematics as a tree, then the Lvov group was devoted to studying roots and trunks, perhaps even the main boughs, with less interest in side branches, leaves and flowers. The deep investigation of the constructions of classical mathematics led the Lvov group to introduce more general notions, which could serve as a basis for broader concepts.

One investigated, from the set theoretic and axiomatic viewpoints, the nature of general spaces rather than specific examples; the general meaning of continuity rather than examples of continuous functions of one variable; the nature of more general sets of points of Euclidean space rather than classical geometric figures; the likely objects of study were general functions of one and several variables, the general notion of curve length, area and volume, i.e., the notion of measure, and the general concept of probability. One studied and compared known mathematical constructs, and common structural features were abstracted from them.

The general results could then be interpreted in each of the specific cases, without the need to devise new proofs for each concrete example. For instance, many well-known mathematical spaces satisfied axioms of what later became known as a Banach space.

In retrospect, it seems a little strange to me that algebraic ideas were not considered in Lvov in a similarly general context. It is clear that the Lvov group was numerically small, and a development of algebra in the modern spirit had to await the creation of other centers in other countries. It is also curious that the study of the foundations of physics, and in particular the study of space-time, has not been undertaken anywhere in this spirit to this day. It is not surprising that in such a general approach new and strange mathematical objects appear along with the general classical ideas. For example, in topology, in parallel with all familiar geometric figures one encounters pathological continua of points of the plane and of the three-dimensional space.[2]

In the study of functions of a real variable, it turned out that nondifferentiable functions form a "majority" among all the continuous functions. In the study of infinite-dimensional vector spaces, it turned out that whole families of them are as important as the Hilbert space. The analysis of various properties of functions, their differentiability and different kinds of continuity led us to understand that each of these notions is connected to a certain infinite-dimensional vector space, often as interesting as the Hilbert space itself. Properties of sequences of real numbers, their convergence and summability have been considered by means of the vector spaces of such sequences. The study of foundations, i.e., the axiomatic formulation of probability theory, demanded an investigation of very general measures and the construction of new spaces of complex "events," which were constructed out of given spaces.

The excitement caused by finding such a great variety of new objects, which one could dissect by means of a few general tools, was so great that the frequency of group discussions in those years was indeed exceptionally high. The only other case of a similar commonality of interests and intensity of intellectual intercourse which I encountered was during the war years when I worked on the then new problem of nuclear energy.

Mathematics in Stefan Banach's Time

by Wojbor A. Woyczyński

AGAINST THE BACKGROUND of steady progress in the great scientific centers of England, France, Germany, Italy, and Russia, three sizzling developments in the last quarter of the 19th century prepared the ground for the massive explosion of new ideas in pure mathematics at the beginning of the 20th century:

- The creation (basically, single-handed) of the *theory of infinite sets* by Georg Cantor (1845–1918);

- Felix Klein's (1849–1925) announcement in 1972 of the *Erlanger Programm* which proposed geometry as a discipline concerned with the study of an abstract object invariant under given transformation groups;

- The appearance in 1899 of *Grundlagen der Geometrie* by David Hilbert (1862–1943) axiomatizing Euclidean geometry.

All three came from Germany. They brought about a fundamental change both in the position of mathematics among other disciplines of knowledge, and the way mathematicians think about themselves. The aftershocks lasted well into the 1930s and beyond.

As a result, mathematics broke away from the body of natural sciences. It embraced the study of arbitrary structures and abandoned its role as truth about the physical world. However, as in the case of Adam and Eve and of Icarus, this new intoxicating liberty carried a heavy price. Mathematics as truth about nature had no problem with consistency – contradictory theorems could not arise; but, following the creation of Riemannian geometry, by the 1880s it became clear that, for example, Euclidean geometry was not true in any absolute sense. The best one could do at the time was to reduce the problem of its consistency to the problem of the consistency of arithmetic. The latter remained a thorn in the side of abstract mathematics for quite a while. In 1900 Hilbert placed it as the second problem on his famous list of twenty outstanding problems. He presented them at the Second International Congress in Paris, and they were meant as a sweeping but concrete agenda for 20th century mathematics.

*   *   *

Cantor, professor at the University of Halle, published between 1874 and 1897 his fundamental and, at the time, totally novel set theory in a series of papers in the *Mathematische Annalen* and the *Journal für Mathematik*. This provided the framework for a study of infinite sets and created a hierarchy of infinities. Denumerable and nondenumerable infinities were quite different. Since he mostly worked with point sets in Euclidean spaces, the study of point set topology was a natural part of his program. The important notions for the analysis of limit points and open and closed sets were then formalized.

It took quite a while before set theory was accepted and absorbed by the mathematics world, and the process was not without controversies and misgivings. Playing fast and loose with set-theoretic notions led to logical paradoxes that called into question the very consistency of mathematics. The flavor of those difficulties can be easily understood through the so-called "barber" paradox: The only barber in town boasts that obviously he does not shave those people who shave themselves, but does shave all those who do not shave themselves. A smart-aleck sidekick challenges him whether he should shave himself. Immediately, the barber finds himself in a logical quandary. If he should shave himself, then by the first part of his statement, he should not shave himself. On the other hand, if he does not shave himself, then by the second part, he must shave himself.

Another antinomy, known as the Richard–Berry paradox, that the set of all integers that can be defined by an English language sentence consisting of less than one hundred letters is finite (because of the finite alphabet). Hence, the complement of this set has the smallest integer. As a result, the definition of "the smallest number which cannot be defined by a sentence containing less than one hundred letters" is contradictory, as it itself contains less than 100 letters.

These were only two of many such paradoxes discovered at that time and, in 1908, in the treatise *Science et méthode*, Henri Poincaré (1854–1912) lamented "How can our intuition deceive us to such a degree?" There were also other philosophical difficulties, mainly engendered by the so-called "axiom of choice," related to the very existence of certain infinite mathematical objects. The ground was fertile for major breakthroughs in research on foundational problems and, as it turned out, some of the most profound discoveries in 20th century mathematics occurred in that area.

* * *

One of the basic questions was whether mathematics could be developed as an extension of logic. The latter had already acquired an axiomatic foundation by 1879 in the work of Gottlob Frege (1848–1925), professor at Jena. This path was followed by Cambridge mathematicians Bertrand Russell (1872–1970) and Alfred North Whitehead (1861–1947); their work culminated in the publication in 1910–1913 of their monumental three-volume *Principia Mathematica*. Paradoxes were avoided by the introduction of the theory of types and mathematics (and logic) followed from the axioms of logic without any axioms of mathematics proper.

Consequently, mathematics had no content or physical meaning – merely form.

Another approach to the foundations of mathematics was offered by Hilbert himself who, from 1904 until the end of his life, *without* relying on set theory, tried to provide a basis for the number system and establish the consistency of arithmetic. The general idea was that mathematics would thus be a collection of many branches, each with its own axiomatics, and with logic formalizing the reasoning process. Axioms were merely rules by which formulas followed from each other. His proof theory, a method of establishing consistency of any formal system, became very influential and Hilbert firmly believed that it would solve the foundational problems in mathematics.

Then came Kurt Gödel's (1906–1992) bombshells. In a 1931 paper in the *Monatshefte für Mathematik und Physik*, he showed that it is impossible to establish the consistency of any formal system containing standard logic and number theory (such as Russell–Whitehead's or Hilbert's) within the system itself. Moreover, he demonstrated the *incompleteness theorem*, which asserted that in any theory of the above type there is a statement such that neither it nor its negation are provable. Since one of them is true, the existence of true but unprovable statements follows. Shellshocked by these discoveries, Hermann Weyl quipped that "God exists since mathematics is consistent, and the devil exists since its consistency cannot be proved."

These soul-wrenching results left in shambles the concept of mathematics as just a collection of axiomatized branches (although, even today, one can find mathematicians who cling to it). However, it did not destroy the value of the axiomatization processes themselves, as they helped to strip away the superficial layers of assumptions of mathematical theories and dug deep in search of their essential ingredients. New theories were blossoming, and the abstract approach was established as the mainstream in pure mathematics. More global and all-encompassing approaches to problems became *de rigueur*. Abstract algebra became independent of such classical notions as real and complex numbers. "Nobody can expel us from the paradise created for us by Cantor!" Hilbert exclaimed in 1930 in the seventh edition of his *Grundlagen*.

Hugo Steinhaus reminisces in his posthumously published *Memoirs and Notes* that Einstein's special relativity theory was considered in Göttingen as a "...triumph of the axiomatic method. Hermann Minkowski presented special relativity at a session of the Mathematical Society giving it a completely mathematical form. He started with praising the greatness of Albert Einstein's scientific achievement, but added that 'the mathematical education of the young physicist was not very solid, which I'm in a good position to evaluate since he obtained it from me in Zurich some time ago.' "

* * *

The emergence of set theory also permitted the creation of a new abstract and general framework for the theory of measure (length, area, volume, etc.). The decisive contribution here was made by the French mathematician Henri Lebesgue (1875–1941). He first presented his ideas in 1902 in the *Annali di Matematica Pura et Applicata*. Thus the foundation for the modern theory of integration and of measuring the size of very complex sets (fractals included) was established. Furthermore, the new measure theory created a renewed impulse for a study of functions.

These new areas did not develop apart from one another. Interactions between them were common and, as a result, yet newer branches which emerged on the borderlines of the older ones began to have lives of their own. Significantly, the process of applying new ideas to classical problems continued. As time passed, the ability to produce such applications was often decisive in evaluating the usefulness of a new branch.

Such was the case with functional analysis. With new achievements in the theory of integration, progress in the study of classical integral functional equations was not far behind. Vito Volterra (1860–1940), professor of mathematical physics at the University of Rome, contributed a basic method for their solution (he also coined the term "functional") but it was again David Hilbert who, in a series of six papers from 1904 to 1910 published in the *Nachrichten von der Königlichen Gesellschaft der Wissenschaften zu Göttingen*, made the seminal contribution. In particular, he rigorously showed how the integral equations can be viewed as infinite-dimensional limits of finite-dimensional systems of algebraic equations. This was a milestone in the development of functional analysis, although it was John von Neumann (1903–1957) who, in two papers in *Mathematische Annalen* in 1929, provided proper axiomatization of what we now know as a Hilbert space and of the operators in a Hilbert space. He also made a crucial observation that Hilbert spaces provided correct formalism for the just-emerging quantum mechanics.

The latter events were preceded by Banach's "seven years of plenty." Between 1922, when he published his concept of complete normed spaces which contained Hilbert spaces as the special case, and 1929, when his systematic study of dual spaces appeared, mature functional analysis emerged. Within that period, three fundamental theorems to which the theory owes much of its power and depth were obtained: the Hahn-Banach theorem on extensions of linear functionals, the Banach-Steinhaus theorem on the sequence of linear operations, and Banach's closed graph theorem. In 1932 these results became the cornerstones of Banach's seminal *Theory of Linear Operations*. The monograph also connected functional analysis with the wealth of its applications and inspired numerous further studies. Decades later, individual chapters of that book grew into broad new disciplines of research. Currently, functional analysis is one of the most active branches of mathematics. Jean Bourgain obtained his 1994 Fields' Medal at the International Congress of Mathematicians in Zurich for contributions in that field.

* * *

The wealth of so many new and abstract ideas and the fact that understanding them, although intellectually demanding, often did not require broad erudition, gave many a talented student of mathematics an equal (or perhaps better) chance to master them in competition with silver-bearded, settled-in-their ways mathematics professors. In this climate, new centers of mathematics emerged where there were none before. A centuries-long tradition of high-level mathematics research was no more a *sine qua non* condition of excellence.

Modern ideas of international mathematics were relatively quickly intercepted by Wacław Sierpiński (1882–1969) and Hugo Dionizy Steinhaus (1887–1972). Both of them had spent time in Göttingen. Steinhaus studied there for five years, and in 1911 obtained his Ph.D. there under David Hilbert's guidance. They began developing new mathematics in Poland by offering university lectures on these topics. Sierpiński's textbook on set theory appeared in 1912 and was one of the first published anywhere. It gave gifted students a quick entry into this cutting-edge research.

Sierpiński settled in Warsaw and Steinhaus in Lvov; thus two main centers of modern mathematics in Poland were created. The ties between the Warsaw and the Lvov groups were very strong as far as the exchange of problems, ideas, and people. Unavoidably, there was also a degree of rivalry between them that, however, had a very positive effect on scientific creativity.

The Warsaw group, centered around Sierpiński and Stefan Mazur-kiewicz (1888–1945), quickly attracted young, talented, and very active mathematicians such as Kazimierz Kuratowski (1896–1980) and Karol Borsuk (1905–1982). Their main interest was topology, and the center quickly achieved international recognition. Some of the younger men in that group emigrated in the 1930s to the United States and became prominent members of the American mathematical community – for example, Nachman Aronszajn at the University of Kansas and Samuel Eilenberg at Columbia. Independently, a group strong in logic and the foundations of mathematics developed in Warsaw around Jan Łukasiewicz (1878–1956), and later Alfred Tarski (at Berkeley after World War II) and Andrzej Mostowski (1913–1975). The commonly used terms *Polish space* in metric topology and *Polish reverse notation* in logic and computer science honor their contributions.

The Lvov group, at least in the beginning, worked on applications of the newly-created measure theory: integral theory, measurable functions, functional series, etc. Banach, attracted to scientific work by Steinhaus, very quickly overshadowed his master, and after a few preparatory papers in which his ideas matured, achieved a synthesis that was very much in the spirit of the era. Instead of a detailed investigation of separate functions, he introduced general objects, which today are called *Banach spaces*, and showed broad possibilities for their applications. He harnessed intuitions about familiar geometric objects and algebraic operations to a study of functions considered as points of spaces of infinite dimension. In a sense, he accomplished an algebraization of infinite-dimensional geometry; that was done for "physical" three-dimensional space by Descartes in the 18th century. Following his approach, the proofs of many results on functions became much simpler, more elegant, and devoid of often tedious and complicated "calculations." Another significant discovery of the Polish group was that of the very broad category method. It demonstrated that sometimes it is easier to show that *most* of the objects have a certain property than to produce a single example of such an object. Initially, some mathematicians working in more classical areas of mathematics displayed distrust of the new, as they said "soft" approach. Eventually their doubts were dispelled, as functional analysis proved its worth by providing powerful tools in many areas of analysis.

Although the impact of Banach's work on the development of functional analysis was decisive, the group of mathematicians assembled by him in Lvov and his inimitable and infectious style of working also played a major role. Z. William Birnbaum, another member of that generation and now Professor Emeritus at the University of Washington in Seattle, reminisces that

> ... Mathematics in this group of interested people was a kind of fever. They were getting together at every possible location, at many different times of the day and night, talking mathematics. There was a large tile stove in a place that was a combination of seminar room and miniature mathematics library. That stove had three sides that were accessible and one that leaned against the wall. I remember hours when people stood around that stove in the cold of night, leaning against those very warm tiles at the three sides that were accessible, and talking around the corner about mathematics. Better known to historians of mathematics are the daily and nightly gatherings of mathematicians in the legendary Scottish Café, Kawiarnia Szkocka.

Stanisław Mazur (1905–1981) made a significant contribution to the process of writing Banach's monograph and developed geometric methods in functional analysis. Juliusz Schauder (1899–1943) was one of the founders of nonlinear functional analysis and topological methods in partial differential equations. Władysław Orlicz introduced and studied functional spaces that are now known under his name. As the European political climate deteriorated before World War II, other young members of the Lvov group also emigrated to America: Steinhaus' student Marek Kac (1914–1984) had a distinguished career in probability and mathematical physics at Cornell and Rockefeller Universities, and Ulam had a significant and well-publicized role in the Manhattan Project and later on the hydrogen bomb.

In the second half of the 20th century, functional analysis exploded well beyond the Lvov school's wildest dreams although the original mainstream – as understood by Banach – continues to survive as a part of the extremely broad area spanned by great discoveries of Laurent Schwartz and Alexandre Grothendieck in France, and I. M. Gelfand[1] and his collaborators in the Soviet Union.

APPENDIX II

Selected Publications of Stefan Banach

ABBREVIATIONS:

BIAP – Bulletin Internationale de l'Académie Polonaise des Sciences et de Lettres, Classe des Sciences Mathématiques et Naturelles, Séries A: Sciences Mathématiques.

CRAS – Comptes Rendus de l'Académie des Sciences (Paris)

FM – Fundamenta Mathematicae

SM – Studia Mathematica

WM – Wiadomości Matematyczne

[1] *Sur la convergence en moyenne de séries de Fourier* (with H. Steinhaus), Bulletin International de l'Academie des Sciences de Cracovie, Année 1918, Classe des Sciences Mathématiques et Naturelles, Séries A: Sciences Mathématiques, pp. 87–96.

[2] *Sur la valeur moyenne des fonctions orthogonales*, BIAP (1919), pp. 66–72.

[3] *Sur l'équation fonctionelle* $f \, x \, C \, y / D \, f \, x / C \, f \, .y /$, FM I (1920), pp.123–124.

[4] *Sur les ensembles de points où la derivée est infinie*, CRAS 173 (1921), pp. 457–459.

[5] *Sur les solutions d'une équation fonctionelle de J.C. Maxwell* (with S. Ruziewicz), BIAP (1922), pp.1–8.

[6] *Sur les fonctions dérivées des fonctions mesurables*, FM 3 (1922), pp. 128–132.

[7] *Sur les opérations dans les ensembles abstraits et leur application aux équations intégrals*, FM 3 (1922), 133–181.

[8] *An example of an orthogonal development whose sum is everywhere different from the developed function*, Proceedings of the London Mathematical Society (2) 21 (1923), pp. 95–97.

[9] *Sur le problème de la mesure*, FM 4 (1923), pp. 7–33.

[10] *Sur un théorème de M. Vitali*, FM 5 (1924), pp. 130–136.

[11] *Sur une classe de fonctions d'ensemble*, FM 6 (1924), pp. 170–188.

[12] *Un théorème sur les transformations biunivoques*, FM 6 (1924), pp. 236–239.

[13] *Sur la décomposition des ensembles de points en partiens respectivement congruentes* (with A. Tarski), FM 6 (1924), pp. 244–277.

[14] *Sur les lignes rectifiables et les surfaces dont l'aire est finie*, FM 7 (1925), pp. 225–236.

[15] *Sur une propriété caractéristique des fonctions orthogonales*, CRAS 180 (1925), pp. 1637–1640.

[16] *Sur le prolongement de certaines fonctionelles linéaires*, Bulletin des Sciences Mathématiques (2) 49 (1925), pp.301–307.

[17] *Sur la convergence presque partout de fonctionelles linéaires*, Bulletin des Sciences Mathématiques (2) 50 (1926), pp.27–32 and 36–43.

[18] *Sur une classe de fonctions continues*, FM 8 (1926), pp. 166–172.

[19] *Sur le principe de la condensation des singularités* (with H. Steinhaus), FM 9 (1927), pp. 50–61.

[20] *Sur certains ensembles de fonctions conduisant aux équations partielles du second ordre*, Mathematische Zeitschrift 27 (1927), pp. 68–75.

[21] *Sur les fonctions absolument continues des fonctions absolument continues* (with S. Saks), FM 11 (1928), pp.113–116.

[22] *Sur les fonctionelles linéaires*, SM 1 (1929), pp.211–216.

[23] *Sur les fonctionelles linéaires II*, SM 1(1929), pp. 223–239.

[24] *Sur une généralisation du problème de la mesure* (with K. Kuratowski), FM 14 (1929), pp. 127–131.

[25] *Differential and Integral Calculus, Volume I*, (in Polish), Zakład Narodowy im. Ossolińskich, Lvov 1929, 294 pp.

[26] *Differential and Integral Calculus, Volume II*, (in Polish), Książnica-Atlas, Lvov 1930, 248 pp.

[27] *Sur la convergence forte dans le champ L^p* (with S. Saks), SM 2 (1930), pp. 51–57.

[28] *Über einige Eigenschaften der lakunären trigonometrischen Reihen*, SM 2 (1930), pp. 207–220.

[29] *Bemerkung zur Arbeit "Über einige Eigenschaften der lakunären trigonometrischen Reihen"*, SM 2 (1930), p. 251.

[30] *Über additive Massfunktionen in abstrakten Mengen*, FM 15 (1930), pp. 97–101.

[31] *Théorème sur les ensembles de première catégorie*, FM 16 (1930), pp. 395–398.

[32] *Über analytisch derstellbare Operationen in abstrakten Räumen*, FM 17 (1931), pp. 283–295.

[33] *Über metrische Gruppen*, SM 3 (1931), pp. 101–113.

[34] *Über die Baire'sche Kategorie gewisser Funktionenmengen*, SM 3 (1931), pp. 174–179.

[35] *Über die Höldersche Bedingung* (with H. Auerbach), SM 3 (1931), pp. 180–184.

[36] *Theory of Operations. Volume I: Linear Operations* (in Polish), Kasa im. Mianowskiego, Warsaw 1931, viii+236 pp.

[37] *Sur les transformations biunivoques*, FM 19 (1932), pp. 10–16.

[38] *Théorie des opérations linéaires*, Mathematical Monographs 1, Warsaw 1932, vii+254 pp.

[39] *Eine Bemerkung über die Konvergenzmengen von Folgen linearen Operationen* (with S. Mazur), SM 4 (1933), pp. 90–94.

[40] *Sur la structure des ensembles linéaires* (with K. Kuratowski), SM 4 (1933), pp. 95–99.

[41] *Zur Theorie der linearen Dimension* (with S. Mazur), SM 4 (1933), pp. 100–112.

[42] *Sur la dimension linéaires des espaces fonctionelles* (with S. Mazur), CRAS 16 (1933), pp. 86–88.

[43] *Sur les séries lacunaires*, BIAP (1933), pp. 149–154.

[44] *Sur la mesure de Haar, Note au livre: S. Saks, Theorie de l'intégrale*, Mathematical Monographs 2, Warsaw 1933, pp. 264–272. English translation *On Haar's measure* appeared as a supplement to the *Theory of the Integral* by S. Saks, Mathematical Monographs 7, Warsaw-Lvov 1937, pp. 314–319.

[45] *Über mehrdeutige stetige Abbildungen*, SM 5 (1934), pp. 174–178.

[46] *Sur un théorème de M. Sierpiński*, FM 25 (1935), pp.5–6.

[47] *Die Theorie der Operationen und ihre Bedeutung für die Analysis*, Proceedings of the International Mathematical Congress, Oslo 1936, pp. 261–268.

[48] *The Lebesgue integral in abstract spaces*, a supplement to *Theory of the Integral* by S. Saks, Mathematical Monographs 7, Warsaw-Lvov 1937, pp. 320–330.

[49] *Über homogene polynome in* (L^2), SM 7 (1938), pp. 36–44.

[50] *Mechanics for Academic Schools, Part I* (in Polish), Mathematical Monographs 8, Warsaw-Lvov-Wilno 1938, vi+234 pp.; *Part II* (in Polish), Mathematical Monographs 9, Warsaw-Lvov-Wilno 1938, pp. 235–556.

[51] *Über das "Loi suprème" von Hoene-Wroński*, BIAP (1939), pp. 1–10.

[52] *Sur la divergence des séries orthogonales*, SM 9 (1940), pp. 139–155.

[53] *Sur la divergence des interpolations*, SM 9 (1940), pp. 156–165.

[54] *Sur la mesures dans les corps indépendants*, Papers of the Institute of Mathematics of the Academy of Sciences of the Ukrainian SSR 8 (1947), pp. 71–90.

[54a] *On measures in independent fields* (edited by S. Hartman), SM 10 (1948), pp.159–177.

[55] *Remarques sur les groupes et les corps metriques* (Prepared for publication by S. Hartman), SM 10 (1948), pp. 178–181.

[56] *Sur les suites d'ensembles excluant l'existence d'une mesure* (Posthumous note with a preface and commentary by E. Marczewski), Colloquium Mathematicum 1 (1948), pp. 103–108.

[57] *Sur la représentation des fonctions indépendantes à l'aide des fonctions de variables distinctes* (Posthumous note edited by S. Hartman and E. Marczewski), Colloquium Mathematicum 1 (1948), pp. 109–121.

[58] *Introduction to the Theory of Real Functions* (in Polish), Mathematical Monographs 17, Warsaw-Wrocław 1951, iv+224 pp.

[59] *Mechanics*, Mathematical Monographs 24, Warsaw-Wrocław 1951.

[60] *Oeuvres*, Volumes I and II, Polish Scientific Publishers, Warsaw 1967.

Textbooks by Banach

[1] *Arithmetic and Geometry for the 5th Grade of Elementary Schools* (with W. Sierpiński and W. Stożek), Lvov-Warsaw, 1933.

[2] *Aithmetic and Geometry for the 1st Grade of Middle Schools* (with W. Sierpiński and W. Stożek), Lvov, 1929.

[3] *Arithmetic and Geometry for the 2nd Grade of Middle Schools* (with W. Sierpiński and W. Stożek), Lvov, 1930.

[4] *Arithmetic and Geometry for the 3rd Grade of Middle Schools* (with W. Sierpiński and W. Stożek), Lvov-Warsaw, 1931.

[5] *Arithmetic for the 1st Grade of Gymnasiums* (with W. Sierpiński and W. Stożek), Lvov-Warsaw, 1933.

[6] *Algebra for the 2nd Grade of Gymnasiums* (with W. Stożek), Lvov-Warsaw, 1934.

Selected Bibliography

T HIS WORK HAS a popular-scientific, or even journalistic character, rather than a strictly scientific one. Hence it does not contain detailed commentaries or precise references. The same applies to the bibliography. I will not bore the reader by listing all my sources, sometimes consisting of a single sentence. Below, I will limit myself to a quick survey of my main sources of information in case someone would like to delve deeper.

A. Basic information can be gleaned from issues of the following newspapers, journals, periodical and series:

(1) *Czerwony Sztandar* [Red Banner], Lvov, years 1940, 1945,

(2) *Gazeta Lwowska* [Lvov Gazette], years 1920, 1941,

(3) *Fundamenta Mathematicae,*

(4) *Kronika Uniwersytetu Jana Kazimierza* [Chronicle of the Jan Kazimierz University],

(5) *Nauka Polska* [Polish Science],

(6) *Mathesis Polska*,

(7) *Monografie Matematyczne* [Mathematical Monographs],

(8) *Prace Matematyczno-Fizyczne* [Mathematical-Physical Papers],

(9) *Robotnik* [The Worker], organ of the Polish Socialist Party, year 1945,

(10) *Rocznik Polskiej Akademii Umiejętności* [Annals of the Polish Academy of Knowledge],

(11) *Rocznik Towarzystwa Naukowego Warszawskiego* [Annals of the Warsaw Scientific Society],

(12) *Studia and Materiał y z Dziejów Nauki Polskiej* [Studies and Materials in the History of Polish Science],

(13) *Studia Mathematica*,

(14) *Wiadomości Matematyczne* [WM, Mathematical News, Annals of the Polish Mathematical Society].

The following articles and reminiscences were most useful in preparation of this book:

- Albiński M., Wspomnienia o Banachu i Wilkoszu [Remembering Banach and Wilkosz], *WM* 19 (1975–76).

- Aleksandrow P.S., List do prof. K. Kuratowskiego [A letter to prof. K. Kuratowski], *WM* 12 (1969).

- Bielajew W., Professor Stefan Banach, *Czerwony Sztandar* 174 (1945).

- Courant R., Wspomnienia z Getyngi [Reminiscences of Göttingen], WM 18 (1974).

- Gołąb S., Z dziejów założenia Polskiego Towarzystwa Matematycznego [From the history of establishing the Polish Mathematical Society], *WM* 4 (1961).

- Gołąb S., O dorobku matematyków polskich w nauce światowej [On the contribution of Polish mathematicians to world science], *Życie Nauki* 13–14 (1947).

- Janiszewski Z., O potrzebach matematyki w Polsce [On needs of mathematics in Poland], *WM* 7 (1963–64).

- Kac M., Henri Lebesgue i polska szkoła matematyczna: obserwacje i wspomnienia [Henri Lebesgue and the Polish mathematical school: observations and reminiscences], *WM* 20 (1976–1978).

- Kuratowski K., Polskie Towarzystwo Matematyczne w okresie międzywojennym [Polish Mathematical Society between two world wars], *Nauka Polska* (1969).

- Leja F., Powstanie Polskiego Towarzystwa Matematycznego [Establishment of the Polish Mathematical Society], *WM* 12 (1971).

- Marczewski E., Rola matematyki w poznawaniu i kształceniu świata zewnętrznego [Role of mathematics in the study and shaping of the external world], *Nauka Polska* (1971).

- Mazur S., Przemówienie wygłoszone na uroczystości ku uczczeniu pamięci Stefana Banacha [An address delivered at the Stefan Banach memorial conference], *WM* 4 (1961).

- Mierzecki H., Po potopie [After the deluge], *Nowa Epoka* 25 (1945).

- Miś B., Opowieści Kięgi Szkockiej [Tales of the Scottish Book], *Perspektywy* 12 (1971).

- Orlicz W., Sur l'oeuvre scientifique de Stefan Banach, *Colloquium Mathematicum* (1948).

- Sieradzki J., Z każdym świętem umieraliśmy [We died on each holiday], *Odrodzenie* 47 (1945).

- Ślebodziński W., Polskie Towarzystwo Matematyczne w latach 1919–1963 [Polish Mathematical Society in the years 1919–1963], *WM* 8 (1965).

- Sierpiński W., Wstęp do: *Towarzystwo Naukowe Warszawskie* [Introduction to *Warsaw Scientific Society*], Warsaw 1932.

– Sobolew S. Ł., Przemówienie wygłoszone na uroczystości ku ucz-czeniu pamięci Stefana Banacha [An address delivered at the Stefan Banach memorial conference], *WM* 4 (1961).

– Stark M., Hugo Steinhaus jako nauczyciel w okresie lwowskim [Hugo Steinhaus as a teacher in the Lvov period], *WM* 17 (1973).

– Steinhaus H., Autobiografia [Autobiography], *WM* 17 (1973).

– Steinhaus H., Przemówienie wygłoszone na uroczystości ku uczcze-niu pamięci Stefana Banacha [An address delivered at the Stefan Banach memorial conference], *WM* 4 (1961).

– Steinhaus H., Souvenir de Stefan Banach, *Colloquium Mathematicum* (1948).

– Stone H.M., Nasz dług wobec Stefana Banacha [Our debt to Stefan Banach], *WM* 4 (1961).

– Szökefalvi-Nagy B., Przemówienie wygłoszone na uroczystości ku uczczeniu pamięci Stefana Banacha [An address delivered at the Stefan Banach memorial conference], *WM* 4 (1961).

– Szpilrajn E., Święto matematyki polskiej [A celebration of Polish mathematics], *Wiadomości Literackie* 3 (1936).

– Ulam S., Przygody matematyka [Adventures of a mathematician], translated by Jerzy Jaruzelski, *Kultura*, July 30, 1978; August 6, 1978; August 13, 1978.

– Ulam S., Wspomnienia z Kawiarni Szkockiej [Reminiscences from the Scottish Café], *WM* 12 (1969/71).

– Warhaftman S., Pierwszy Polski Zjazd Matematyczny [First Polish Mathematical Congress], *Mathesis Polska* 2 (1927).

– Wojciechowski E., Profesor dr. Rudolf Weigl, *Służba Zdrowia* 35 (1957).

– Zygmund A., Międzynarodowy Kongress Matematyczny w Oslo [International Mathematical Congress in Oslo], *Mathesis Polska* 11 (1938).

– Zygmund A., Otwieranie drzwi [Cracking the door open], A conver-sation with Professor Antoni Zygmund (by Jerzy Jaruzelski), *Kultura*, October 22, 1978.

❖ *Appendix III* ❖

B. Books:

- Colerus E., *Od tabliczki do różniczki,* [From multiplication table to derivatives], translated by A. Nukliński, introduction by S. Banach, Lvov 1938.
- Dianni J., Wachułka A., *Tysiąc lat polskiej myśli matematycznej* [A thousand years of Polish mathematical thought], Warsaw 1963.
- Dorabialska A., *Jeszcze jedno życie* [Yet another life], Warsaw 1973.
- Estreicher K., *Leon Chwistek. Biografia artysty* [Leon Chwistek. Biography of an artist], Warsaw 1979.
- Gołembowicz W., *Uczeni w anegdocie* [Scholars in anectode], Warsaw 1973.
- Kieniewicz S., *Historia Polski 1795–1918* [History of Poland 1795–1918], Warsaw 1968.
- Kuratowski K., *Notatki do autobiografii* [Notes for an autobiography], Warsaw 1981.
- Kuratowski K., *Half a century of Polish mathematics*, Warsaw 1973.
- Mauldin R.D., Editor, *The Scottish Book. Mathematics from the Scottish Café*, Birkhäuser 1981.
- *Okupacja i ruch oporu w Dzienniku Hansa Franka 1939–1945* [Occupation and the resistance movement in the Hans Frank's Diary], Warsaw 1970.
- Orłowicz M., Ilustrowany przewodnik po Lwowie [The illustrated guide to the city of Lvov], Lvov-Warsaw 1925.
- Sawyer W., *Droga do matematyki współczesnej* [Path to modern mathematics], Warsaw 1969.
- Steinhaus H., *Wspomnienia* [Reminiscences], Cracow 1970.
- Struik D., *Krótki zarys historii matematyki do końca XIX wieku* [A brief outline of the history of mathematics untill the end of XIX century], Warszawa 1960.
- Ulam S., *Adventures of a Mathematician*, New York 1976.
- Wiener N., *I am a Mathematician*, London 1956.

126

– *Wkład polaków do nauki* [Contribution of Poles to science], J. Hurwic, Editor, Warsaw 1967.

C. Original sources:

– Archives of the Metropolitan Archidiocese of Cracow

– Records of the St. Lazarus Parish in Cracow

– Records of the St. Nicolaus Parish in Cracow

– Archives of the St. Szczepan Hospital

– Archives of the town of Jordanów (near Myślenice), XVIII–XX centuries

– Archives of the Health Office of the city of Cracow 1841–1951

– The Great Address Book for the Royal Capital City of Cracow, 1892–1920

– Census of the inhabitants of the city of Cracow 1892–1920

– Cracow Calendar by Józef Czech

– Cracow Calendar for 1892

– Reports of the Principal of the Imperial-Royal H. Sienkiewicz IV Gymnasium in Cracow

– Schematics of the Kingdom of Galicia and Lodomeria with the Grand Duchy of Cracow, 1914

– Archives of the Jagiellonian University

– Archives of the Employment Agency of the City Magistrate

– Curricula of the Imperial-Royal Polytechnical School in Lvov

– The State Archives in Cracow

– Proceedings of the Scientific Society in Lvov, years 1920–1939

– 1936 Polish Mathematical Society questionnaires, edited by K. Kuratowski

– Lvov Address Register of merchants, industrialists and professions

Notes

Chapter 1

1 Kałuża notes that St. Lazarus still exists today, and that aside from being a fine medical institution, the hospital has been the birthplace of many distinguished Polish men and women who have made their mark in the arts, sciences, politics, and so on.

2 The primary doctor on duty in the ward at the time of birth was Maurycy Madurowicz, assisted by doctor Kazimierz Smorągiewicz, and the midwife Magdalena Przybyłowa.

3 No one else seems to have any information about her, although there are speculations about her coming from an aristocratic background.

4 Steinhaus' short biographical sketch of Banach is the only one known to me.

5 This was the Austrian Imperial-Royal Main Tax Office, located at 23 Basztowa Street, since the area of Poland around Cracow and Lvov was then part of the Habsburg Empire.

6 The Gazette was the organ of the [Austrian] Imperial-Royal governor; hence the information was reliable, having come from the court

journal which published various legal and administrative announcements.

7 Some 50 km south of Cracow.

8 A circular park surrounding the center of Cracow, it was planted in the 19th century after the city moat was filled in with earth

9 Gymnasium is equivalent to high school.

10 Okocim beer is still available in fine gourmet shops in the U.S. and Europe.

11 In Europe the *matura* is the final high school graduation examination, required for admission to university entrance examinations.

12 Religion: 2 hours per week. History of the Old Order [Old Testament].

Latin: 8 hours per week. Continuation from the first grade program: conjugation of regular and irregular verbs, declensions of nouns, syntax of simple sentences; oral and mnemotechnical [memory] exercises, as in the first grade; three school assignments every month and one home assignment.

Polish: 3 hours per week. Grammar: review of material covered in the first grade – compound sentences, types of subordinate clauses; spelling and punctuation; read poetry and excerpts from literature aloud; stylistic essays three times a month, alternating school and home assignments.

German: 5 hours a week. Conversation in the form of questions and answers based on fragments read earlier, memorizing words, phrases, and whole prose fragments; review of the regular declensions and main principles of syntax; an assignment every week, including one home assignment per month.

History and Geography: 4 hours per week. History of the ancient world, especially of Greece and Rome, using the biographical method; physical and political geography of Asia and Africa; vertical and horizontal topography of Europe; a detailed course in the geography of southern Europe and Great Britain; exercises in making cartographical sketches [maps].

Mathematics: 3 hours per week. Review and complements on the greatest common divisor and the least common multiplicity; systematic study of regular fractions; conversion of regular fractions into decimal fractions and vice versa; relations, proportions; the rule of "three singles" with application to proportions; inference, calculus of simple interest. From geometry: axes of symmetry of intervals and angles, congruent triangles and their applications. Most important properties of the circle, quadrangles, and polygons. Exercises and problems, as in the first grade.

Natural history: 2 hours per week. For the first six months: birds, reptiles, amphibians, fishes, crustaceans, worms, molluscs, echinoderms, and protozoans. From March on, the world of plants.

13 The ranking system was part of the elaborate career ladder that civil servants had to climb under the Austrian bureaucracy. The VIIIth rank was quite high.

14 Jan Jaglarz and Michal Bogucki, both of the VIIth rank, taught Greek and the latter, Latin as well; Stanisław Koprowicz gave Polish language instruction, while Jan Kreiner, Ph.D. in Philosophy, did the same for German, Karol Stach for Greek, and Gustaw Tellier for French; and Wladyslaw Kudelka taught natural history.

15 In his biography of Chwistek.

16 Bronisław Malinowski, a well-known ethnographer and one of the pioneers of sociology in the United States, and Stanisław Witkiewicz (Witkacy), the well-known playwright and painter, expressed similar sentiments.

17 According to Albiński's memoirs.

Chapter 2

1 From a conversation between Włodzimierz Stożek and Andrzej Turowicz, as reported to the author by Turowicz.

2 *Habilitation* or *Venis legendi* – the right to give lectures independently. It used to be, and in some countries still is, a major step in scientific careers in Europe.

3 The highest level was Professor Ordinarius, despite what is suggested by the titles.

4 From Professor Alicja Dorabialska's memoirs, *Yet Another Life.*

5 Ibid.

6 This was confirmed by Stanisław Ulam, a distinguished mathematician and a member of the Manhattan Project at Los Alamos during World War II, who mentioned Banach's early youth in his *Reminiscences from the Scottish Café.*

7 We have to draw conclusions on the basis of general information, since nobody who knew Banach in those days responded to the author's newspaper appeals.

8 From Turowicz's conversations with Kałuża.

9 From Janiszewski's article (1918).

Chapter 3

1 Mathematical-Physical Papers.

2 The University was created in 1661 by King Jan Kazimierz who issued a decree that transformed the existing Jesuit College into the Lvov Academy. During the partition of Poland between Russia, Prussia, and Austria, that is, from 1794–1918, the University was an important center for science and education for all Polish patriots. Before World War I, it bore the name "The Imperial-Royal Francis I University," and from 1918–1939 it was called Universytet Jana Kazimierza (Jan Kazimierz University). The University building at 4 St. Nicolaus Street, a huge edifice built in the style of an Austrian military barracks, belonged at one time to the Samuel Głowiński Foundation. After restoration of the Polish state, a significant number of university departments were moved to the former legislature building (at 1 Marszałkowska Street) which, from then on, was popularly nicknamed "the new university." Banach obtained his habilitation from the Philosophical Faculty, headed by Dean Kazimierz Kwietniewski and Associate Dean Zygmunt Weyberg. Besides the

Faculty of Philosophy, the University was comprised of the Faculties of Theology, Law, and Political Knowledge, and Medicine. The poet Rector Jan Kasprowicz then headed the university.

Chapter 4

1 Based on Steinhaus' memoirs.

2 *Über analytisch darstellbare Operationen in abstrakten Räumen* – on operations that have analytic representations in abstract spaces; *Über metrische gruppen* – on metric groups; and *Über die Baire'sche Kategorie gewisser Funktionenmengen* – on Baire category of certain sets of functions; Über die Höldersche Bedingung—on Hölder property.

Chapter 5

1 The Warsaw mathematician Edward Marczewski was a measure theorist and algebraist. During World War II he was sent to perform slave labor in the Silesian city of Breslau. There he survived a long Russian siege. The city did not fall to the Russians until May 6, 1945, just a few days before Germany's surrender. Marczewski settled in Breslau after the war and became Rector of the Polish University (now Wrocław University) and a leader of the new mathematical school that arose from the joint efforts of former Warsaw and Lvov mathematicians.

2 Mariacki, Halicki and Bernardyński are among the lovely public areas mentioned by the author

3 In *The Scottish Book: Mathematics from the Scottish Café*, R.D. Mauldin, Editor, Birkhäuser 1981

4 These are approximate counts as some problems were not dated.

5 Author's footnote: Beginning in 1935, Ulam spent the academic year at Princeton, and later at Harvard, but always returned to Poland for his summer vacation.

6 He goes on to say: "I still remember how, in January of 1920 – I guess, waiting for a train at the railway station in Lvov, where he had relatives, he caught a cold, came down with pneumonia and soon thereafter died."

Chapter 6

1 Mazur, on the other hand, was in the 1930s an active member of the Polish Communist Party, although, according to Turowicz, nobody at the Scottish Café seemed to have known about it. Following the war, Mazur became a high official in the new communist-led science establishment. Given how close they were, it seems unlikely that Banach did not know about Mazur's political involvement.

2 Author's footnote: Professor Alexiewicz, who was in touch with Banach during the Soviet and German occupations of Lvov disputes this claim and states that Banach's evaluations of political situations always turned out to be correct.

3 Here, Steinhaus is trying to reproduce the kind of earthy way in which Banach often expressed himself.

Chapter 8

1 Author's note: from Stefan Banach's introduction to the Polish translation of the book *From Multiplication to Differentiation* by Egmont Colerus.

2 Including recently popularized fractals.

Appendix I

1 *Functional Analysis on the Eve of the 21st Century: In Honor of I. M. Gelfand*, S. Gindikin, J. Lepowsky, R. L. Wilson, editors, Progress in Mathematics 131, 132, Birkhäuser, Boston, Basel, Berlin, 1995.

Index of Names